INTEGRATING TECHNOLOGY IN HIGHER EDUCATION

Edited by

M. O. Thirunarayanan
Aixa Pérez-Prado

University Press of America,® Inc.
Lanham · Boulder · New York · Toronto · Oxford

4501 Forbes Boulevard
Suite 200
Lanham, Maryland 20706
UPA Acquisitions Department (301) 459-3366

PO Box 317
Oxford
OX2 9RU, UK

Printed in the United States of America
British Library Cataloging in Publication Information Available

Library of Congress Control Number: 2004117925
ISBN 978-0-7618-3127-3

The paper used in this publication meets the minimum requirements of American National Standard for Information Sciences—Permanence of Paper for Printed Library Materials, ANSI Z39.48—1984

Table of Contents

Introduction

M.O. Thirunarayanan
Aixa Perez-Prado
Florida International University

Writing an introduction that provides an overview of each of the chapters in a book without stealing the authors' thunder is a difficult thing to do. In these few pages we will not focus exclusively on the specific content of each chapter, but on the major themes and ideas that crisscross the twenty chapters that are a part of this book. The thirty-seven authors represented span four continents and a variety of disciplines. Each chapter offers a unique perspective, and collectively the chapter authors provide a diversity of ideas and experiences for integrating technology in higher education settings.

Four primary themes that form the basis for integrating technology in higher education are either explicitly or implicitly embedded in many of the chapters in this book. These four themes are *infrastructure*, *instructional design*, *integration* and *interaction*.

Infrastructure consists not only of the wiring, hardware and software but also the training and support for faculty and students who utilize such hardware and software for purposes of teaching and learning. It is obvious that faculty who teach in colleges and universities that do not have the necessary hardware and software components of technological infrastructure in place will indeed find it difficult if not impossible to integrate technology into their teaching. Barriers to inclusion of technology within the curriculum include a lack of timely and appropriate support where and when needed (Butler and Sellbom, 2002).

According to Burton, Moore and Magliaro (1996) "Instructional design is generally considered to be a systematic process that uses tenants of learning theories to plan and present instruction or instructional sequences" (p. 58). Using modern technologies without proper planning will most likely result in frustration. Instructional design can help plan and implement technologies in more effective ways. A number of the chapters in this volume describe the need for proper planning in order to effectively integrate technology into higher education classrooms.

The term 'integration' when used in conjunction with technology suggests that technology is utilized as an integral part of teaching and learning. Suggestions for integrating technology are offered by many chapter authors.

Bannan-Ritland (2002) states that "There is consensus among theorists, researchers, and practitioners of educational technology and distance education that interactivity is a critical variable in learning" (p.161). Bain (2004) notes that the best college teachers create environments that allow students to "work collaboratively with others" (p. 18). Interaction, both between students and instructors and among students, is a theme that permeates many chapters in this book. Interactions among students can take a variety of forms including discussions, interactive presentations, and collaboration on course related projects. Technology facilitates both interaction and collaboration among students. Such communications can be either synchronous or asynchronous in nature. Several chapters in this book describe collaboration and discussion among students within the classroom as well across countries and continents. Using portable computer technology to develop an in-class community of learners is also discussed in at least one of the chapters in this book.

The authors of the twenty chapters describe the use of a variety of software and hardware technologies in their classrooms. These include presentation software, spreadsheets, databases, discipline specific software packages, idea organizers, and a visualization tool. In addition to handheld computers or personal digital assistants, Tablet PCs and other computers, authors also discuss the use of videos, simulations and the live broadcast of a lecture without the use of a video camera. The technology use discussed in this book is not limited to just computer hardware and software, but also includes the use of a human patient simulator.

Authors wrote the chapters for this book based on their own technology integration experiences. They provide authentic information based on what they tried, what worked, and what did not work. They talk about issues that arose and the problems that they faced and how these were resolved and solved. Others who wish to integrate technology in their own educational settings can benefit from the tried and tested suggestions offered by the authors of the chapters in this book.

References

Bain, Ken. (2004). *What the best college teachers do*. Cambridge, Massachusetts: Harvard University Press.

Bannan-Ritland, Brenda. (2002). Computer-mediated communication, elearning, and interactivity: A review of the research. *The Quarterly Review of Distance Education*, *3*(2), 161-179.

Burton, John K., Moore, David M., and Magliaro, Susan G. (1996). Behaviorism and instructional technology. (pp. 46-73). In David H. Jonassen. (Ed.). *Handbook of research for educational communications and technology*. New York: Macmillan LIBRARY Reference USA, Simon & Schuster Macmillan.

Butler, Darrell, L., & Sellbom, Martin. (2002). Barriers to adopting technology for teaching and learning. *Educause Quarterly*, *25*(2), 22-28.

Teaching with Technology: Infrastructure Support and Online Discussions

Steven Hornik
University of Central Florida

Julie A.B. Cagle
Xavier University

Abstract

Two faculty members' experiences of incorporating classroom technology as a supplement to traditional FTF classes and to conduct a distance class are described. As instructors we experienced feelings of loss of control. An instructor can plan and design the course content and assignments, but the technology used and infrastructure to support it are largely outside the instructor's control. We offer suggestions on how instructors can gauge infrastructure support at their institutions. Difficulty maintaining participation in online discussions was another common aspect to our experiences. Research suggests students perceive a qualitative difference between FTF and online discussions and technology has yet to bridge this gap. Throughout the paper lessons we learned are laid out to help both faculty and administrators consider key issues that may allow for a positive experience when introducing technology into higher education.

Introduction

There is a burgeoning literature related to lessons learned from technology-enhanced teaching and online classes (Chizmar & Williams, 1998; Cook, 2000; Cook, 2000; McEwen, 2001; and Vinaja & Raisinghani, 2001). Contributing to that literature, this paper reports two faculty members' experiences of incorporating classroom technology and online discussion in their classes. We incorporated components of Technology-Mediated Learning (TML) into MBA and undergraduate classes in different ways: (1) as a supplement to the traditional classroom; and (2) a method of delivery of a distance learning class.

Alavi and Leidner (2001) define TML as an "environment in which the learner's interactions with learning materials (readings, assignments, exercises, etc.), peers, and/or instructors are mediated through advanced information technologies" (p. 2). As instructors, we differed in our technological backgrounds and disciplines, one coming from an Information Systems department and the other from Finance. While content, level, and method of delivery of our courses differed, two aspects of our experiences were very similar. In particular, we both struggled with loss of control when technology didn't work as expected, and with generating and maintaining consistent participation in online discussions. Prior to addressing these two common aspects of our experiences in teaching with technology, we will highlight the role of expectations in the TML experience.

Incorporating Technology into the Classroom: Expectations

The incorporation of educational technology into the classroom is new to most faculty and their students. As a result, there is a greater chance for unrealistic expectations. The lack of experience, and frequent unrealistic expectations are consistent with the high withdrawal rates reported for online courses (Phipps and Merisotis, 2001; and Priest, 2000), as well as the recent development of courses to orient students to online learning environments (Mendels, 1999).

Acceptance of the technology by students is an important issue, particularly for classes that use TML techniques to augment the classroom rather than as part of an online environment. Students can be very apprehensive, and even resistant, about the inclusion of technology to supplement a traditional face-to-face (FTF) class (Arbaugh, 2000). Arbaugh & Duray (2001) have not found any significant relationship between student acceptance of the Information Technology (IT) in TML and perceptions of learning. Student apprehension can take one of two forms: (1) fear of the IT, as evidenced by this typical question; "why do we need to use this technology if we are meeting FTF?" or (2) frustration with the additional task of learning the course material as well as the TML interface (Hillman, 1994), as evidenced by this typical question: "How is this technology going to help me learn accounting?" These are tricky questions that colleagues and administrators might also be asking. Disturbingly, one of the authors received an e-mail outlining the plight of one instructor who was terminated for this very reason. "I was terminated because I dared use some technology in my course! The charge was teaching technology over basics, whatever "basics" means in a first accounting course" (Merwin, 2001).

For the next few years, it is understandable that some students will question the necessity and additional time required to master the use of the classroom technology when they already know it is going to take a considerable amount of time to master the course material. The principal way to help these students is to make the classroom technology supportive of their learning by designing the technology such that: (1) communication with the instructor and with other students is easy (Arbaugh, 2000), (2) self-assessments (e.g., quizzes) are provided, and (3) feedback on assignments is given as soon as possible. The classroom technology chosen by an instructor should not place too much of a burden on the students in terms of learning how to use it, and the technology must supplement the course objectives rather than just be a tool for its own sake.

Instructors must communicate to students what they can expect in terms of feedback from the instructor. Students' expectations of the speed of response from instructors seem to increase with a class that incorporates technology (Cagle & Hornik, 2000). For example a student might post a question at 10:00 p.m. and expect a response quickly enough to incorporate the information into the assignment due by midnight. An instructor needs to communicate to the class the time frame in which they can expect a response and the implications of that time frame for the student's timetable of completing an assignment.

The potential for unrealistic expectations isn't limited to students and faculty. Much to administrators' chagrin, there is increasing evidence that online instruction is less efficient and more costly than traditional instruction. After a year-long study of online education in a faculty seminar (University of Illinois, 1999), the final report describes the shift from the classroom to online teaching as a shift from "efficiency to quality" and smaller student to faculty ratios. Boettcher (1998) suggests that the maximum number of students for online courses is in the range of 12 to 20, depending on the level of instruction. Feenberg (1989) reports that student/faculty ratios for the executive programs like the University of Phoenix are even fewer, at about 8 to 1.

While the smaller suggested class size and need for technological support differentiate online from traditional FTF education, there is a growing pedagogy of online learning (Webster & Hackley, 1997) and more evidence that "good teaching is good teaching" regardless of the delivery mechanism. Recent studies of Internet delivered courses have demonstrated similar results. Peterman (2000) found online courses are successful when the professor is focused on the needs and concerns of the students. Browning (2000), studying the differences between web-delivered and traditional classroom instruction, reports that there was no difference in learning between the two groups in concept understanding. Kirby and Driscoll (1997) postulate that there should be no difference between traditional and distance learning outcomes and suggest that the same factors that affect student learning in a traditional classroom should also affect student learning in a distance class. Arbaugh and Duray (2000) support this argument with their finding that pedagogical, rather than technical factors were most strongly associated with learning in a distance-learning environment. Our similar experiences as faculty using technology both as a supplement to a traditional FTF class and for a distance education course are consistent with this view. The principles that describe a good educational experience are similar regardless of the delivery mechanism. In particular, the development of a relationship between instructor and student is highlighted in the University of Illinois (1999) report as a critical element for a quality online course, along with a caution on class size in order to achieve that necessary rapport. Our experiences affirm the role of "human touch" in distance and traditional classes. However, this acknowledgement merely heightens the role of technology as it provides a means for instructor-student contact, and may in fact be the sole means of such contact in the case of online courses.

Loss of Control and Technological Infrastructure

While a critical element for a quality technology enabled course, the technological infrastructure is generally not within an individual instructor's control. In a traditional classroom, the "sage on the stage" by-and-large determines the outcomes of a traditional classroom experience. In an online class, the "guide on the side" determines the design of the instructional strategy and activities, but not the servers hosting the software, the networks used to access the material, the computers used by students, or the speeds at which students are connecting to the Internet. All these links create a bridge from students to the "classroom." One purpose of this paper is to highlight the need to consider the principles of a good educational experience together with the instructors' power to achieve them when incorporating technology in the classroom.

First, appropriate technology must be in place, and supported, at your institution. The important lesson is that, while administrators may be supportive of instructors adopting new technology, they may not know the condition of the technology within the institution and may not understand the considerable strain these types of technology will place on existing systems. Therefore, if administrators do not have an equal commitment to acquiring the proper hardware, software, and peopleware, faculty members who incorporate technology in their courses might experience high levels of frustration and helplessness.

The course support software used previously at Xavier University was Lotus Notes LearningSpace, and it has now been replaced by Blackboard. With Lotus Notes, individual institutions are responsible for purchasing servers and providing technical support. This requires not only a substantial commitment in terms of investment in hardware, but also human resources. Expertise related to the software application is needed, as well as additional technical support personnel for faculty and students using the software.

One of the major problems that can occur with the adoption of technology is a bit paradoxical: the faster faculty adopts a platform (e.g., LearningSpace or Blackboard), the worse the technological support may become, unless scalability (i.e. hardware capacity) is built into the support budget. This places a heavy burden on the administrative entities of a university to ensure that the knowledge of a particular software package exists or is easily obtainable with training or hiring, and that the existing hardware and support personnel on campus are capable of the increased needs of users. If all these things are not in place, hardships for both the instructor and the student will ensue. At worst, this can lead to a negative backlash against any new technological advances in pedagogy by instructors. One software package might be used, but abandoned in a short period of time as adoption increases the need for unplanned resources. This may occur just as early adopters are becoming proficient in using the software and erase any return from the investment in the software by the early adopters. After these types of difficulties, the early adopters may be reluctant to make the necessary investments required to adopt replacement software packages. Whether the early adopters switch to the new software or not, there will be costs for training the faculty on the use of the new package. Such experiences may also scare off adoption by instructors who are not early adopters.

Unfortunately, most if not all of the infrastructure issues like adequate hardware and technical support are outside the instructor's control. Simply put, faculty members no longer act alone when they enter a classroom. They are a part of an entire delivery system, most of which is invisible to our students. Goodyear, Salmon and Spector (2001) list 8 different roles associated with an online class: 1) process facilitator, 2) adviser-counselor, 3) assessor, 4) researcher, 5) content facilitator, 6) technologist, 7) designer, and 8) manager-administrator. At some institutions, various individuals will assume some of these roles. As a result, faculty must be prepared to give up the control that they are accustomed to having when they "enter" the classroom. Being prepared with backups for the invariable mishaps, is surely important and something that all faculty must do. Having foreknowledge of the loss of control does not in itself help instructors retain control, rather it allows for better contingency planning.

Time and Technological Infrastructure

Teaching online takes more time (Chizmar & Williams, 1998; Eastman & Swift, 2001). The instructor spends more time training and coaching students in online techniques and "netiquette," participating in and monitoring online discussions, and counseling the students in the course as they adapt to a new pedagogy. Palloff and Keith (1999), comparing the time commitment for an online versus a FTF class, found the online class takes up to three times as much time as the FTF class.

We have found that the quality of the technological infrastructure is a contributing factor to the extra time commitment of an online class. The instructor spends substantial time on the technological issues, particularly when the campus technical support is already overwhelmed, since the instructor is the first line of defense when students encounter any technological problem and will be contacted first. This "overhead" time is nothing new. Holmberg and Bakshi (1992) report that time spent on administrative details and acting as a general liaison were 43% and 40%

respectively, while the time spent on course content was only 13% for telephone tutoring.

In some cases, the instructor is also dealing with Information Support Services, or its equivalent, of the student's employer to try to resolve technological issues associated with, for example, firewalls and set-up of Internet browsers. Faculty incorporating technology into classes may, de facto, become the technology experts as well as content experts (Goodyear, 2001; Gibson, 2001). The instructor often becomes the advocate for administrative support for Information Support Services. Indeed, faculty may need to take on a technologist role and be well informed on such issues as: (1) understanding the capabilities and limitations of the available technology platform and infrastructure, (2) diagnosing learners' technical issues and challenges, and (3) using technology at an operational level (Goodyear, 2001).

Online Discussions

As previously mentioned, Lotus Notes LearningSpace was used as the platform for an online course, as well as a support tool for traditional FTF classes. Similar to other course products, LearningSpace offers scheduling, posting of personal and/or contact information to share with other students in the class and the instructor, storage of all digital course materials, assessments, and an online Course Room database that allows asynchronous threaded discussions. Instructors have to adapt to the online discussion environment. "Due to the new context where educational communication is mediated through technology across distance, and normal response cues from students can be delayed or eliminated, faculty need to learn new skills to be effective" (Thach & Murphy, 1995 p. 59). Spector and Teja (2001) suggest that these competencies include: 1) allowing learners' time for reflection, 2) keeping discussions alive and on a productive path, and 3) archiving and organizing discussions to be used in subsequent asynchronous discussions. They (Spector and Teja, 2001) also suggest other competencies for synchronous discussions: 1) establishing ground rules for chat, 2) animating interactions with minimal instructor intervention, 3) sensing how online text messages may appear to distant learners, and 4) understanding the impact of cultural differences.

Using threaded discussions in a class can take several forms. Having students respond to pre-posted questions before class primes the pump for in class discussions. In addition, students can be asked to complete One Minute Papers (Hornik, 1999) after class to provide feedback to the instructor, and between students, on problem areas from previous lectures. Both of these uses of threaded discussions are consistent with the notion of student centered learning (Leidner & Jarvenpaa, 1995), complementing instructor directed learning, and if successful can enrich the FTF experience. However, it has been our experience that the actual amount of student-to-student responses has been inconsistent at best. In order to improve the interactions among students, we have employed various techniques with varying levels of success.

Instructor Pre-Populates the Discussion Database

Pre-populating the discussion database with questions and topics has been found to improve the organization and ease of use of the database, as well as the level of student participation in discussions, but coaching is also needed. Students have to be regularly reminded to use the threading mechanism or they may just post every comment as a new discussion thread. Similarly, students may post a question that has already been posted and answered. The first time teaching a particular course, identifying the topics/questions for pre-populating the database can be difficult. For subsequent course offerings, the transcript of previous online discussions can be used to identify and/or modify the pre-posted topics and questions.

Instructor Feedback to the Students

Students need constant evidence that the instructor is reading their comments or both the quantity and quality of discussion responses fall off quickly. Several methods for doing this have been employed, but the silver bullet has not yet been found. One method involved identifying quality responses and copying them into a special area that students can examine to see what the instructor thought was exceptional. Good responses have been read aloud by the instructor during class meetings for the same purpose, as well as to prime the pump for further in-class discussions. The "favorites" function in LearningSpace to create bookmark type of links of the best responses has also been utilized. Common to all of these techniques is the considerable amount of instructor time needed, beyond the normal time required to read and respond to the comments. Unfortunately, none of the techniques have been completely successful.

In one semester in particular, participation levels in the online discussions were unusually low. Students complained that the speed at which their commands were processed was too slow. This problem was the result of increasing the number of student users without the addition of more servers, and as previously mentioned, was subsequently solved by adding new servers. This example highlights the loss of instructor control in a technology-enabled course and the need for sustained commitment from administration.

Online vs. FTF Discussions

Even though students are provided with consistent guidelines for participation in the online discussions, participation levels have varied considerably. For a small subset of students, it is clear that they are not comfortable posting typed responses to the database. There is a concern that when others read their question or comment, they will be perceived as "stupid." The lack of FTF interaction with the other students in an online class and/or greater permanence of posted versus verbal discussion comments may contribute to this fear. On the other hand, online discussions may improve the thoughtfulness of comments and increase the level of participation for other groups of students, in particular those communicating in a second language. However, because students have more time to formulate responses, there may be a loss of spontaneity.

In one study that surveyed students as to differences between online and FTF discussion experiences, the results indicate students enjoy the convenience of asynchronous online discussions, and did not perceive lack of spontaneity or visual clues to be significant problems (Tiene, 2000). However, the students responding to the survey did meet FTF once a week. Results may differ for students in online discussions that do not meet FTF. Further, the students preferred the use of the written word because it provided a transcript of the discussion that can be read as many times as necessary. Paradoxically, students indicated they did not prefer online to FTF discussions, and that online discussions should be used as a complement, rather than a substitute, for FTF discussions. The results of this survey are consistent with our own experiences. While online discussions certainly have their advantages, they still do not seem to be perfect substitutes for FTF discussions.

The ease of use, or lack thereof, of the discussion tools should not be underestimated, for no matter how much prompting and/or incentives students are given, if the software makes posting or reading discussion comments difficult, the students will not use the tools. In a study by Arbaugh (2000), perceived interaction variables (e.g., ease of interaction, classroom dynamics, and instructor emphasis on interaction) were all associated with perceived learning. The results (Arbaugh, 2000) suggest that instructors play an important role in enhancing learning in Internet-based courses by facilitating interactions. However, as important or perhaps more important might be the software. If the ease of interaction is low as in the Arbaugh (2000) study, then no matter how much the instructor tries to emphasize interaction, the software can make it difficult to attain. For example, when LearningSpace is delivered to students via an Internet browser interface, and browsers are not set appropriately, it is impossible for the student to tell which postings are new. Students have to read through all discussion threads to find anything new to respond to. In this type of environment, the postings become more of a one on one discussion between the professor and each of the students rather than the hoped for threading of student-student discourse. When new software (Blackboard) was used which included the feature of notification of new posts, the hoped for student-student threaded discussion took place.

Synchronous Discussions

Another type of discussion is the synchronous discussion that takes place in a chat room. Depending on the objectives of a course a chat room environment could be useful for discussing certain issues and for interjecting the spontaneity of thought that may be missing from the asynchronous threaded discussion. Like everything else in the online realm, the technology seems to lag behind the vision and the tools available today can create frustrating experiences for the student and instructor.

The most serious limitation encountered with a chat environment is the number of individuals that can "talk" to each other at the same time, ideally around six to eight people. MBA classes of thirty-five or more are far from ideal. If everyone in class participates, it is impossible to keep up with the discussions. If a large class is split up into smaller groups, the instructor must attend to each group, perhaps doubling or tripling the time needed to cover a topic. This was the option chosen for the online MBA course. The students were given the option of attending one of two chat sessions given for each topic. Even with this splitting of the class, there were times when so many students showed up for a particular chat that those students present were split again. In these situations, two simultaneous chats were conducted and simultaneous threads had to be read and responded to by the instructor. Beyond the additional time, the experience in the smaller group chats can qualitatively differ based on the different personalities involved, and can make uniform assessment of all students in the class problematic.

Another obstacle to a chat session is the inevitability of having students "talking" over one another, as well as the instructor. It takes on average two chats before the students get comfortable with reading each others' thoughts as they scroll through their computer monitors, first assimilating the information, and then responding. Many times multiple topics are being discussed simultaneously, which points out one potential fallacy of synchronous discussion, that is, a single idea can be discussed sequentially unlike the start-stop nature of an asynchronous discussion.

The technical capabilities of a chat tool are important. The tool packaged with Blackboard was overly simple (e.g., it was slow and without the capability for participants to identify which comments were submitted by the instructor and which comments were made by the students), so an external tool was substituted. Yahoo! Instant Messenger and Yahoo! Chat provided the ability to change the color of the text to help differentiate the instructor's comments from the students' postings. In addition, they provided emoticons that can help overcome limitations imposed by the lack of visual cues in a chat environment.

One problem that can occur when the chat tool is external to the campus software is the potential for individuals not enrolled in the course to enter the chat room and interject comments that are wholly unrelated to the discussion, even if the chat has been made private. Another concern is that chat room environments are not necessarily stable. Because of this instability, a chat would sometimes have to be stopped and restarted mid-chat, resulting in a choppy discussion. As with the asynchronous discussions, much is needed and expected for chat sessions to be successful, from instructor preparation and training in this new fast paced environment, to the capabilities of the software and student understanding of its use, and, as always, a reliance on the "magical" software and hardware to work as expected. While a chat allows spontaneity, which may be lacking in the asynchronous environment, the speed at which ideas are scrolling and in which responses have to be posted precludes deliberative and contemplative responses.

As with asynchronous discussion, research suggests some difficulty in explaining participation in chats. Rollman (2000) examined the effects of a number of variables on reciprocal conversation in three Yahoo portal chat rooms. Neither perceived gender nor general versus specific content of the messages affected reciprocal communication (Rollman, 2000). However, the length of written statements and the provocativeness/rudeness of the comment were found to increase reciprocal communication (Rollman, 2000). While an instructor may well advise students regarding the length of their messages, they may want to avoid rudeness or provocative content in messages given the academic nature of the chats.

Need for Flexibility

Flexibility on the part of the instructor is also important if classes involve significant amounts of online discussions. Servers will go down and students will miss participating in discussions for reasons beyond their control or experience. An online instructor has to be ready and willing to listen to the circumstances and make a judgment call when it is deemed necessary. For example, the first few weeks of the course are when students learn that their corporate firewalls may prevent them from posting discussion topics. Similarly, a student may travel to a country in which they do not have Internet access.

Students must also be flexible in an online course. Taking an online course is a new experience for most students and they have to adapt to self-directed learning, as well as use new techniques like threaded discussions and synchronous discussions. It is similarly important for the educational program to be flexible. Online education may not be the best alternative for every student. For a student that is an auditory learner, a course with a significant online discussion component may be problematic. Students at Xavier have the choice of FTF or online delivery of a course.

Conclusion

Adopting educational technology into your courses can be successful, but like all good teaching takes a lot of time and effort. The lessons we learned can be divided into things in the instructor's control and things outside the instructor's control. For example, instructors should use technology only if it supports course objectives, enables them to communicate to students what they can expect in terms of feedback, and helps develop a relationship with each student. These are in the instructor's control and can be handled via the design of the course.

However, even with instructor commitment there are still challenges to be met in terms of online discussions. Research regarding online versus FTF discussions is still in its early stages. While the evidence suggests there is something different about the two forms of discussion those differences have not yet been identified, nor has a means to address such difference pedagogically been developed. Future research should attempt to discern what is so compelling about FTF that it cannot yet be replaced by online discussions, and then software must be developed to bridge this gap. Until that time, courses that rely heavily on class discussion should include FTF time.

Another aspect critical to the success of a technology-enhanced course that is outside the instructor's control is the technological infrastructure. We offer this caution: no matter how much time and effort an instructor is willing to invest, support structures must also be in place to ensure success. While the instructor can largely influence the degree of human interaction, feedback, and relationship between technology and course objectives; the technological infrastructure is generally the result of sustained organizational and administrative commitment. Evidence of such commitment should be sought before an instructor invests the considerable time and resources required to successfully incorporate educational technology into their teaching. An on-going commitment to faculty development in the area of educational technology is an example of such evidence. However, administrators themselves may not recognize the current or future adequacy of technology and support given the speed of change in the use of technology. Faculty should consult early adopters on campus for the best information about the administrative resources necessary for faculty support once they make the jump to teaching with technology.

To summarize, we have found the following to be helpful in creating a successful online experience, whether you are using TML as a supplement to an FTF course or to deliver totally online classes:

- Make sure the technology used makes sense for the course objectives.
- Use the technology to provide feedback and self-assessments to the students.
- Provide feedback to the students as soon as possible and within the time frame they have been told to expect.
- Determine your institution's commitment to hardware, software and support personnel.
- Become competent in the delivery system and expect to provide help-desk type answers.
- Be willing to take on multiple roles, as suggested by Spector and Teja (2001) and Goodyear (2001), beyond the "sage on the stage."
- Pre-populate the discussion database with questions.
- Provide frequent evidence to students that the instructor is in fact reading ALL discussion posts.
- Use discussion technology that provides the capability to easily navigate to newly posted discussions.
- Use multiple, simultaneous chats to keep chat groups small in size.
- Learn to effectively manage small chat room groups.
- Learn to utilize chat room functionality that differentiates the instructor chat from the students (using color, different font, etc.).
- Learn when to appropriately utilize emoticons.

We have found technology supported teaching to be pedagogically effective, as well as rewarding; though it is not without costs, especially in terms of time commitment and frustration associated with loss of control. It is our hope that those pursuing this new and exciting form of teaching will benefit from our experiences.

References

Alavi, M. & Leidner, D. E. (2001). Research commentary: Technology mediated learning–A call for greater depth and breadth of research. *Information Systems Research*, *12*, 1-10.

Arbaugh, J.B. & Duray, R. (2001). Technology acceptance, perceived flexibility, and student learning and satisfaction with Internet-based courses: An explanatory comparison of two on-line MBA programs. *Academy of Management Best Papers Proceedings*. D. Nagao (ED.), MED A1-A6.

Arbaugh, J. B. (2000). How classroom environment and student engagement affect learning in Internet-based MBA courses. *Business Communication Quarterly*, *63*, 9-26.

Boettcher, J. (1998). How many students are "Just Right" in a web course? *Syllabus Magazine*, *12*, 1.

Browning, J. (2000). *Analysis of concepts and skills acquisition differences between Web-delivered and classroom-delivered undergraduate instruction technology courses*. Unpublished dissertation, North Carolina State University.

Cagle, J. & Hornik, S. (2000). *Using the Internet to teach: Lessons learned from two experiences in a small liberal arts university*. Paper presented at the Stop Surfing Start Teaching: Teaching and Learning through the Internet 2000 National Conference, Charleston, S.C.

Chizmar, J. F. & Williams, D. B. (1998). Internet delivery of instruction: Issues of best teaching practice, administrative hurdles, and old-fashioned politics. *Campus - Wide Information Systems*, *15*, 164-173.

Cook, K. C. (2000). Online professional communication: Pedagogy, instructional design, and student preference in Internet-based distance education. *Business Communications Quarterly*, *63*, 106-110.

Eastman, J. & Swift, C. (2001). New Horizons in distance education: The online learner-centered marketing class. *Journal of Marketing Education*, *23*, 25-34.

Feenberg, A. (1989). The written word: On the theory and practice of computer conferencing. In R. Mason & A. Kaye (Eds.) *Mindweave: Communication, Computers, and Distance Education*, pp. 22-39. Oxford: Pergamon Press.

Gibson, J. W., Tesone, D. V. & Blackwell, C. (2001). The Journey to Cyberspace: Reflections from three online business professors. *SAM Advanced Management Journal*, *66*(1), 30-34.

Goodyear, P., Salmon, G. & Spector, M. J. (2001). Competencies for online teaching: a special report. *Educational Technology Research and Development*, *49*, 65-72.

Hillman, D. C. A., Willis, D. J. & Gunawardena, C. N. (1994). Learner-Interface interaction in Distance Education: An Extension of Contemporary Models and Strategies for Practitioners. *The American Journal of Distance Education*, *8*, 30-42.

Holmberg, R. & Bakshi, T. (1992). Postmortem on a distance education course: Successes and failures. *The American Journal of Distance Education*, *6*, 27-39.

Hornik, S. (1999). *Implications of the electronic one minute paper*. Paper presented at the Association for the Advancement of Computing in Education, Seattle, WA.

Kirby, E. & Driscoll, M. (1997). *Facilitator and student roles and performance in a high school distance education course*. Paper presented at the American Education Research Association Annual Meeting, Chicago, Illinois.

Leidner, D. E. & Jarvenpaa, S. L. (1995). The use of information technology to enhance management school education: A theoretical view. *MIS Quarterly*, *19*, 265-291.

McEwen (2001). Web-assisted and on-line learning. *Business Communication Quarterly*, *64*, 98-103.

Mendels, P. Courses that teach how to learn online. (Saturday/1999, October 16) *New York Times on the Web*.

Merwin, K. (2001). How important is a class syllabus. *AECM archives* - October 2001, week 3.

Palloff, R.M. & Keith., P. (1999). *Building learning communities in cyberspace*. San Francisco: Jossey-Bass.

Peterman, T. (2000). Elements of success at a traditional/virtual university: lessons learned from three years of growth in cyberspace. *Journal of Academic Librarianship*, *26*, 27-32.

Phipps, R. & Merisotis, J. (1999). *What's the difference? A review of contemporary research on the effectiveness of distance learning in higher education*. Washington, D.C.: Institute for Higher Education Policy.

Rollman, J., Krug, K. & Parente, F. (2000). The chat room phenomenon: reciprocal communication in cyberspace. *CyberPsychology and Behavior*, *3*, 161-166.

Spector, J.M., & de la Teja, I, (2001, December). *Competencies for online teaching*. ERIC Digest EDO–IR-2001-09. Syracuse, NY: ERIC Information Technology Clearinghouse. Available at http://www.ericit.org/digests/EDO–IR-2001-09.shtml

Thach, E. C. & Murphy, K. (1995). Competencies for Distance Education Professionals. *Educational Technology Research & Development*, *43*, 57-79.

Tiene, D. (2000). Online discussions: A survey of advantages and disadvantages compared to face-to-face discussions. *Journal of Educational Multimedia and Hypermedia*, *9*, 371-384.

University of Illinois. (1999). *Teaching at an Internet Distance: the Pedagogy of Online Teaching and Learning*. University of Illinois Faculty Seminar. Retrieved August 11, 2003, from http://www.vpaa.uillinois.edu/reports_retreats/tid.asp?bch=1

Vinaja, R. & Raisinghani, M.S. (2001). *Analysis of strategies used in teaching an online course in a predominantly Hispanic university*. Paper presented at The Consortium for Computing in Small Colleges, Branson, Missouri.

Webster, J. & Hackley, P. (1997). Teaching effectiveness in technology-mediated distance learning. *Academy of Management Journal*, *40*, 1282-1309.

By Instructional Design: Facilitating Effective Teaching and Learning with Technology

Donna Harp Ziegenfuss
Widener University

Abstract

Many changing and evolving facets of higher education have contributed to the emergence of new opportunities for teaching and learning. With these new opportunities for change also come new responsibilities for faculty and the necessity to learn new skills as they integrate new technology tools into their teaching. This chapter supports the view that an understanding of instructional design principles and practices can help faculty members organize and more efficiently utilize many new tools, as well as the massive amounts of information that have recently become available for the practice of teaching and learning. Focus will be on describing a process for planning course design and technology integration and alleviating the apprehension faculty may have about how to do this and where to start.

Introduction

It should come as no surprise to anyone associated with an institution of higher learning that teaching and learning on the college campus is not what it used to be. Complicated and interwoven issues such as changing student demographics, technological advancements, the emergence of for-profit educational institutions, stricter financial constraints, and a movement to improve student outcomes and assessments have all contributed to a paradigm shift that is taking place today on many college and university campuses (Schuyler, 1997). Futurist, Alvin Toffler (1991) identifies a shift from an industrial age to an age of information and describes three waves of change that have occurred historically over time. Reigeluth (1996) also makes distinction between the new "information-age paradigm" of instructional theories and the old "industrial-age paradigm" of instructional theories. Barr and Tagg (1995) refer to the shift in the educational world as moving from an "Instruction" paradigm to a "Learning" paradigm. This chapter will provide a brief introduction to some of these contributing change factors associated with this shifting educational landscape, but the primarily focus will be on new opportunities and alternatives made available to faculty as a result of these changes.

Although many factors have contributed to the remaking of this new higher education learning environment, the most influential factors are the introduction of the computer and the explosion of Internet resources into the higher education classroom. These technological advancements have provided unlimited access to global resources and information, for both faculty and students that were not available before, or even imagined of being available before. However, just having access to more information does not automatically make for better instruction. Faculty quickly came to realize the drawbacks of easily accessible information. New issues and problems arose resulting from the rapid deployment of Internet technologies and the need for new skill sets for filtering and evaluating web information. Piggybacking on this issue of excessive resources is the fact that this newly available pool of information is also in a state of constant flux. Learning how to keep up with information updates, additions, deletions, and changes can in itself become a full time job. Time constraints have now become a common complaint of many faculty members who are struggling to keep up with the new advancements in technology and the explosion of information (King, 2003).

These new technological advancements have also created a new breed of college students. Life and professional priorities of current students are different from past generations of students (O'Banion, 1995; Sacks, 1998; Tapscott, 1998). Student demographics have changed and more adult students are visible in the higher education classroom as they change careers, update skills and knowledge, and/or return to the working world. Many students are working and attending classes concurrently, and similar to the college faculty, time has also become one of their top priority issues (Lankard, 1997; Tapscott, 1998). Students have also become more extrinsically motivated when it comes to getting an education. It is not about learning for the sake of learning anymore, and they put a high value on their time spent in class. Time constraints bring students to expect instantaneous gratification and the rapidly changing and improving technology tools have created a student that expects speed, convenience, and entertaining educational multimedia. TVs, DVDs, MTV, streaming web video and digital cameras all contribute to the fact that students are more visual learners than they were a decade ago. Tapscott (1998) describes "'N-Geners' or 'The Net-Generation' as the first generation to grow up surrounded by digital media" (p. 1). He describes N-Geners as interactive, curious, self-reliant, determined, smart, focused, adaptable, high in self-esteem, and possessing a global orientation. They have come to expect web services that enable them to register online, get grades online, and buy books online. They want course content material to be available on a 24/7 basis in the same way they expect their money to be available 24/7 with their ATM cards. They have predefined goals that require specific professional skills, training, and knowledge for employment promotion and come into the higher education learning environment with different ideas and expectations than students did in past decades.

This push for flexibility and convenience, the changes in student population demographics, and the availability of technological advancements have also facilitated an emergence of a different type of learning environment. Web-enhanced and totally online courses are becoming commonplace on many campuses around the country. Teaching and learning in cyberspace requires a different mindset and a different set of tools than teaching and learning in a traditional face-to-face learning environment. Faculty are faced with having to develop new strategies, skills, and new technological proficiencies on top of their traditional responsibilities associated with designing and teaching a college level course. Fullan (2000, p. 16) states, "Technology generates a glut of information, but it has no particular pedagogical wisdom." He (Fullan, 2000, p. 16) concludes by adding that, "This means that teachers must become experts in pedagogical design."

Many faculty members could understandably associate this educational paradigm shift with a reason to panic. Mezirow (1991), a leading researcher in the area of transformational change and adult learning, contends that a "disorienting dilemma," can prompt a more critical reflection on the dilemma or event and facilitate the development of new ways of looking at and learning from our experiences. From an instructional design perspective, I concur with Mezirow (1991) and prefer to view these unsettling technological advancements as an opportunity to reflect on course goals and objectives, redesign course frameworks, and revaluate how we assess student learning in the higher education classroom. This can be an opportunity to try out the new technologies, methodologies and ideologies in teaching. It can be a time to refocus on how students learn rather than on how they are taught. We can use this opportunity to examine and evaluate how the new tools, materials, and learning environments can facilitate more effective teaching & learning. Watts (2003) claims that:

> Technology is also a catalyst for change because in our struggle to master its use, we have by necessity become learners. This affords us with the opportunity to be in touch with how it feels to be on the other end of the teaching/learning spectrum. (p. 6)

Without the development of skills for designing a solid pedagogical foundation on which to build a course or redesign learning, all these new fangled technology tools, information, resources, and learning environments could be rendered useless. The most significant opportunities for a transformation of teaching and learning in higher education could lie in the reflection on teaching and learning and thinking more about the instructional design of the course and student learning outcomes. Brown (2000) states in a book he edited spotlighting best practices in teaching with technology, "Suddenly, with computers, today's professors have a bafflement of alternatives. Hard choices must be made" (p. 3). Understanding the basic principles and processes of instructional design can become a beneficial and essential tool for faculty as they struggle with making choices and decisions concerning the glut of opportunities now available to them.

What is Instructional Design?

Instructional design (ID) is the process of helping students to learn by analyzing all the various variables that impact the learning experience. According to Reigeluth (1983), instruction is any method that can help people learn. The goal of instructional design is to facilitate more effective and meaningful student learning. It is a process of facilitating the cognitive learning processes by linking solid learning theory with relevant and active instructional practices. Instructional design is a holistic process that considers ALL of the components of the teaching and learning experience such as prior knowledge and experience, learner needs and styles, the type of learning environment, the content or learning materials, affective factors, as well as activities and interactions supporting the learning of course content. ID is a process or a way of thinking about the systems and components involved in the learning process and draws on the research and knowledge available about how people learn.

ID is a process for planning that includes (a) identification of learner needs or problems, (b) design of learning materials and instruction to meet the needs of the learners, (c) integration of teaching and learning strategies appropriate to the learners' needs, and (d) an analysis and assessment of the learning outcomes. It is not a once and done deal and requires continuous assessing and adjusting to make sure that teaching and learning goals and objectives are attained. The process and theories of ID are changing and evolving just as other educational variables and theories change. Traditional ID theory, originally based in behaviorist pedagogy, is often criticized for being too linear and new research is developing in the domain of ID as related to online learning, multimedia development and socio-cultural learning environments (Gros et al., 1997; Jonassen & Rohrer-Murphy, 1997; Reigeluth, 2002; Wilson, 1997; and Winn, 1997). ID guides and directs the practice of instructional designers and plays an important and necessary role in the design and development of instruction.

What does ID have to do with the integration of technology into higher education, you ask? ID strategies and methodologies can be effectively utilized in many different types of instructional planning situations. The planning, organization, and pedagogical issues associated with teaching and learning effectiveness become even more important when technology tools are introduced into the teaching and learning process. Technology should not be used just for the sake of using technology. Rather, technology should be used to enhance solid instructional practices or provide an instructional component that could not have been achieved previously without the use of technology.

Pouring a solid foundation of good pedagogical design before adding on the layer of technology can become a critical factor in the success rate of technology integration. ID and technology have been successfully linked in the literature and researchers have documented that integrating technology tools can facilitate successful student-centered active learning experiences (Grabinger, 1996; Jonassen, 1996; and Pascarella & Terenzini, 1991). Learning theories and instructional design models such as constructivism, problem-based learning, activity learning, case studies, and situated learning have become more prominent in the teaching and learning environments of today (Brown, 2000; Brown, Collins & Duguid, 1989; Hannafin, Land, & Oliver, 1999; Jonassen, 1991; and Schon, 1991). These various learning methodologies and models suggest alternatives to traditional passive lecture teaching methodologies and suggest strategies whereby learners become participants in the learning process and actively construct knowledge. Technology tools provide additional strategy choices for instruction that can facilitate communication, collaboration, construction of knowledge and the attainment of course goals and objectives. However, the technology tools must be integrated and utilized appropriately. Brown (2000) states, "More computers equal more learning. This rule is broken by occasional neglect, misuse, or mismatch, but overall there is a positive relationship between the availability/use of computers and the amount of learning that occurs" (p. 4). The newness factor associated with technology tool integration warrants close inspection and documentation so that faculty have scholarly research and examples on which to base decisions before attempting to integrate technology.

Learning instructional design strategies give faculty yet another way to enhance and improve the planning, implementation, and evaluation of learning outcomes in the traditional classroom, as well as in technology-related situations. Each teaching and learning situation is unique and therefore requires individualized planning, implementation and evaluation strategies. Understanding the instructional design process will help faculty make sound decisions for a course redesign and assessment of student learning and teaching styles. In some cases, technology may not be the best tool for the job. As an instructional designer I will be the first to recommend a more traditional instructional methodology to a faculty member if I feel it would do a better job at attaining course goals than a technology-based solution. Technology just offers faculty more strategies and alternatives for solving problems or redesigning a course. Understanding the relationship between technology tools, teaching strategies and effective learning will provide faculty with better insight to judge whether or not technology is the appropriate tool for achieving the predefined learning goals and objectives.

Stages of Instructional Design

Although discipline and content vary significantly from field to field, there are some basic principles of ID and popular ID models that can be adapted to many different situations and instructors. Common traditional ID models such as the ADDIE model (ADDIE stands for Analyze, Design, Develop, Implement and Evaluate), the Dick and Carey model (1978), and the Kemp et. al., model (1997) are three of several models available that can be used by faculty to help design courses based on sound instructional design theory. Each has a slightly different approach to the design problem, yet each model strives to identify a problem or conduct a needs assessment, advocates the setting of goals and objectives, focuses on the designing of learning activities or a means to attain those goals, and then finishes the process by assessing whether the goals were met.

The main difference between the five-step process outlined in this chapter and other published models of ID is that I emphasize a differentiation of the pedagogical planning from the learning activity, tool, or method planning. I contend that in the case of technology integration planning, it is important to establish and define pedagogical goals before beginning to focus on the application of technology tools. At first glance this process will appear to be common sense. However, the power of the ID process is the conscious recognition and reflection on the steps and their sequential implementation as a structured design plan when designing instruction or developing technology-enhanced instruction. The planning process must be made explicit. This step-by-step process of instructional design is one that I have found to be effective in planning my own teaching strategies, as well as a method I recommend when I am wearing my instructional designer hat and helping other faculty redesign instruction or develop new instructional strategies. These instructional design strategies can be applied when developing a technology-related unit or project, designing an entire course, or even when evaluating and developing instructional programs.

Step 1: Needs Assessment and Analysis of Learners

In the needs assessment step of ID, the focus should be on determining what students have coming into the course and what they ideally should have when they leave your course. In this stage of ID it is also important to keep in mind the institutional culture, as well as the program and departmental goals. It is wise to make sure a course design fits into the big institutional picture and builds on prerequisite courses, and also fulfills the requirements for the courses that will follow it. The broad course goals of a project, unit, or module are set in this stage of planning.

Some of the questions to ask in this stage of ID are (a) what type of learners are you dealing with? (adult vs. traditional), (b) how will this type of student learn best? (collaborative projects, individualized learning, active learning, reflective learning), (c) what motivates your particular type of learner? (intrinsic or extrinsic motivation, learning styles, personality types), (d) what kind of prior knowledge do the students have coming into your course? (e) how technology savvy are the students? and (f) why are they learning (why are they in your class)?

As an example of the types of questions I ask and the process I go through during Step 1 of ID, I will discuss some specific issues related to a course I have recently redesigned and teach titled, "Image Processing in Multimedia Design." The course is a required core course for a graduate program for K-12 teacher candidates enrolled in an Instructional Technology Specialist Certification or a Masters Degree program in Educational Technology or Instructional Design. My student population is comprised of adult learners working as teachers or in instructional/technological positions and taking coursework part time. I consider their special needs as adult learners such as flexibility, desire for relevant hands-on, and authentic instruction, and a need to design their own learning based on prior knowledge. These learner characteristics are typical for adult learners as found in the research of Brookfield (1986), Knowles (1980), Lawler (1991), and Mezirow (1991).

In the course design I provide hands-on group instruction and also offer choices to design and complete individual projects that are relevant to their professional experiences. In many cases, school districts are subsidizing the tuition for these students and demand an A or B grade in order for the student to be eligible for tuition reimbursement which acts as an external motivator for attaining good grades. On the other hand, these students are intrinsically motivated by their interest in educational technology and this is their reason for being enrolled in the technology and education program in the first place. I conduct an online pre-course survey where I ask questions about their interests, prior experience, motivation, and expectations so I can be better prepared to customize course activities and content.

Students coming into this course, which focuses on the development of digital graphics and animations, have taken a prerequisite course in web design. After meeting with the department coordinator and the instructor teaching the prerequisite web course, I have decided to use the development of a web-based portfolio as the capstone project for my course. This requirement ensures that students develop additional web-based skills in this course, as well as build on prior skill sets and knowledge developed in the prerequisite course. Providing an opportunity for students to practice again and review prior knowledge is a successful strategy that perhaps also facilitates improved knowledge retention. The students are very technology savvy, since that is the purpose of the program, and they are expected to have access to and experience with a range of technology tools.

This course is delivered in a hybrid format (partially online and partially face-to-face) to facilitate the adult learners' need for flexible schedules. This class meets five times during a traditional semester of fifteen weeks with substantial work and assignments also required in an online modality. This format requires an additional dimension in planning, since I must decide which learning activities can be completed most successfully in an online or distance modality and which learning activities are best handled in the face-to-face environment.

In setting broad goals for this course, my main goal is to help students understand what multimedia development is, how to use software tools to develop multimedia projects, and most importantly to understand the learning theory and instructional design principles implicit in effective multimedia design. I expect that when the course is completed, student-learning outcomes will include having:

- learned the technology skills required to develop effective multimedia
- conducted a literature review and written a short scholarly research paper on a multimedia learning topic that is of interest to them and relevant to their professional careers
- actually developed a multimedia portfolio.

Step 2: Design of the Course Pedagogical Framework (Detailing the Goals)

In this step of the instructional design process, sub-goals and/or objectives are written and linked to the broad course goals developed in Step 1. This step includes the development of the syllabus, or the specific plan for the semester, and delineates what is to be done and in what order it is to done to achieve the course goals. You might think of this step in the planning process as deconstructing or dissecting your large course goals into manageable objectives or chunks of sub-goals and making sure that student learning can be demonstrated and measured for each of the objective chunks. It is important to think of it as focusing in on or drilling down into your course as you move through the ID process steps. A demonstration of learning may be performance-based such as demonstrating the skills to develop a multimedia portfolio or knowledge-based such as researching a topic, synthesizing information, drawing a conclusion and writing a paper to demonstrate what has been learned.

I think that this step is the most important step in ID. If the specific course objectives are not succinct and do not relate directly back to the course goals, then there is no accurate way to design learning activities that demonstrate student knowledge. For example, one of my main goals from Step 1 is for students to learn the principles of effective graphic design and multimedia development. I need to define how I will know if they have learned the principles of multimedia design in order to be able to measure learning. Do they need to memorize the names of Mayer's (2000) eight principles of multimedia design? Do they need to develop an eye for appealing graphic design? How do they demonstrate that they understand and can apply design concepts? In this case I set objectives or sub-goals that further define my expectations for learning. Some objectives might include (a) students will be able to distinguish between good and bad design samples, (b) students will be able to identify the elements of a good design, (c) students will read and research a topic related to multimedia design and share what they learned with fellow students to clarify their understanding of graphic and multimedia design, and (d) students will be able to apply the principles of design to their professional situations and create effective multimedia materials for teaching and learning.

Each of these objectives supports the main goal of learning the principles of effective multimedia design. If students can demonstrate mastery of these objectives, then they can also demonstrate that they have achieved the main learning goal. In this second step, we are still planning pedagogical goals and objectives and there is still no mention of how technology could be used to achieve these goals. It is easier to envision possible teaching and learning strategies first and clarify expectations for learning before adding the technology component. If feasible goals and objectives are not established in the first two steps of the ID process, then throwing good technology tools at poor goals and objectives will not facilitate more effective learning. Weak course goals will produce uncertain objectives and uncertain objectives will result in learning outcomes that cannot be demonstrated.

Step 3: Designing Interactive Components for Attaining Goals and Objectives

Steps 1 and 2 focus on the course content and application of the content, or in other words, what will students learn, when will they learn it, and how will they learn it. How will students apply knowledge to demonstrate they have learned the content? In this step of the ID process, the focus narrows in on the learning environment, the teaching strategies, and the teaching tools become the priority concern. Learning activities, student interactions, tasks and/or skills are then assigned to each of the specific objectives so that knowledge and mastery of the subject can be demonstrated.

Before deciding on specific learning activities or projects consider the "Cone of Experience" classifications of Dale (1969) that advocate using a variety of instructional strategies such as text, audio/media, simulations or hands-on experience. Any of these types of learning strategies can be used successfully to achieve course or technology integration objectives and goals. One course may actually incorporate a variety of these strategies and methodologies to accommodate the wide variety of learner types. Some additional questions to ask when beginning to define which tool or strategy is right for the job are (a) will there be individual projects or group projects? (b) will there be a variety of instructional media used in the course (videos, audio, animations, web-based products or text)? (c) will there be lecture-style teaching or facilitator style teaching? (d) will there be study groups, collaborative projects, or individualized projects?

See Table 2.1 to view some possible learning activities and related teaching strategies associated with one of the objectives that I used as an example in step 2 of the ID process. The objective I refer to is that students will be able to distinguish between good and bad design samples. I have considered three different modalities for achieving this objective. Then possible learning activities or methods for each modality are listed, as well as possible (technology-based and non-technology based) alternatives. Notice that the face-to-face (F2F) lecture and demonstration option could be achieved by using a technology tool like PowerPoint, or a non-technology option such as the traditional F2F discussion. I did not start out planning learning activities by selecting PowerPoint as the teaching strategy and then trying to figure out how I would use it to achieve the objective. Rather, I designed the learning activity to use first, which was to use authentic printed and web-based examples to teach the principles of good design, and then began to think about which tools or strategy would best achieve those goals.

Step 4: Design and Development of Instructional Materials or Modules

Once the decisions are made on which learning activities and tools would most effectively implement the course objectives, it is time to begin the development of the instructional materials. In the case of technology-related projects, it may be necessary to design and produce a PowerPoint presentation, or set up online discussion forums using a course management software tool. A technology-based strategy may also involve digitizing or scanning in materials or images. There is preparation involved even if you decide to use a non-technological teaching and learning approach, such as going through magazines and collecting design samples to share in class.

The preparation and development step must be even more seriously considered when redesigning instruction or developing technology-related materials. A common design and development mistake that I see, as an instructional designer, is a lack of consideration for the amount of time required to "develop" the course design plan. Teaching/learning activities are planned but not enough time is allocated to develop the materials required to carry out the teaching strategy.

Table 2.1. Planning Learning Activities and Teaching Strategies

Objective: Students will be able to distinguish between good and bad design samples.

Modality Options	Types of teaching/learning activities	Tool and/or approach options
F2F Lecture & Demonstration	Use lecture to introduce and demonstrate the graphic design and multimedia learning principles through examples. Supplement lecture content with actual design examples, ads, and web resources as examples to visually represent principles and concepts.	Could use a PowerPoint presentation to organize principles and examples in a single lecture presentation. Could use hard copy printouts or magazine page examples that could be passed around and discussed in small groups. Using real examples introduces an authentic learning component to the lecture.
Online self-directed activity supplemented with F2F follow-up	Students would search for design examples in magazines and on the web (one good and one bad) and post their critique about the designs online. The online critique experience and discussion will be continued in the F2F class meeting	The choice of good and bad examples and critique will demonstrate the student understanding of and application of design knowledge. The critique process will facilitate critical and analytical thinking and stimulate both online and FTF student discussion and thinking. F2F sessions can be reserved for confirming student progress and filling in content not covered online.
Online self-directed activity	Design examples will be provided for the students to review online. Critique process will be modeled by the instructor. Students will participate in discussion of the critiques. Students will then submit their own designs online for critique. Online class discussion will guide student's redesign possibilities.	Practice with visual examples and the critique process in an online format will prepare them for preparation and posting of their own designs in an online format. Preliminary design work can be done online and the F2F sessions can then be used for trouble-shooting problems and individual assistance.

I recommend to faculty that as they begin to integrate technology into their courses, that they keep track of the amount of time that it takes to scan in images, or create PowerPoint presentations, or populate online courses with content. This time management strategy will enable them to judge the amount of time they will need to do specific tasks for future projects and will facilitate more efficient course planning in the future.

The development step becomes a much more prominent ID step when technology is integrated into courses. It is wise to plan extra time dedicated to the development of technology materials but it is also wise to plan for "technology gone bad" time. The first time you develop or use a technology strategy it is safe to assume that you may have to make adjustments or rethink your strategy after you have started developing. So always allow even MORE time than you think you will need to develop and integrate new technology-related materials.

Step 5: Implementation, Assessment and Adjustment.

The final step in the ID process involves the implementation, assessment and then readjustment of the teaching and learning strategies. In this stage you will plan how you will assess what you have built or developed. Several ID models look at the implementation and assessment stages as two separate steps, but I prefer to think of these two steps as one. I recommend that you develop the habit of implementing and assessing concurrently. All of the steps in this process can actually be concurrent and there may be a need, for example, to circle around and return to planning when developing. Assessment conducted during a course or when testing a course is called formative assessment and can be a valuable tool in fine tuning and reworking strategies as you go along. I like to think of formative assessment as "in the trenches" assessment. It is best not to wait until the end of the semester to find out what went wrong. When the assessment process is part of the design and implementation process you can adjust on the fly and involve students in the process of evaluation. Formative assessment practices may involve conducting a student online or traditional paper survey several weeks into the course to find out how things are going. By making a concerted effort to collect data throughout your course, you will be forced to be in tune with what is really going on. Data collected during and after the course can also become valuable institutional and departmental outcome data that may in the future provide evidence that could be used to justify additional technology related expenses, staffing and resources.

Questions you might ask students during this last step in the ID process are (a) what do they like about the course? (b) what do they not like about the course? (c) what is their favorite part of the course? (d) if you are trying a technology-based strategy, ask them about what technical concerns or issues they encountered. Are they having any technical problems? (e) do they prefer one teaching strategy or modality over another for a particular concept? (f) how did they feel about participating in and discussing topics online?

Don't be afraid to ask. When students realize that you are seriously using their feedback to make teaching and learning adjustments I have found they can be candid and reflective. Also do not be afraid to admit that something did not work. When something does not work you will have two choices, one to try to fine tune the approach to remedy the situation, or the other to completely readjust your strategy to include another approach altogether. Keeping a journal of your observations and reflections about the course and student learning as you go through the process will also be valuable and authentic information that you can read though after the course is finished and look for patterns, successes and places for improvement that can be incorporated in your next project.

In addition to "real-time" assessment you must also design summative assessment strategies that you will use at the end of the course to evaluate student-learning outcomes. The evaluation and assessment plan, similar to the learning activities and strategies, must also be tied to the course goals and objectives. I recommend to faculty that they design a survey instrument or assessment strategy in the beginning of the course, since it is best to design an assessment strategy when you are first designing the course goals and objectives. There will always be things that come up during the course that you could not have anticipated and you may need to refine your assessment strategies at a later date, but do not wait until the end of the course to think about how you will assess student learning.

Summary and Conclusions

The right tools and strategies for the job become even more important in this time of many choices. Decisions about what tool to use, when to use it, and how to use it are critical questions that must be considered when designing for successful instruction. Included here are some guidelines and suggestions that I regularly offer to faculty contemplating a change in their teaching strategy or when planning a technology integration project.

Ten ID Suggestions and Guidelines for Technology Integration

1. Instructional design before technology.

In higher education, the process of course planning and emphasis on course goals and objectives has not changed over many years, but the methods that can be used to attain those goals have changed dramatically over the years. Course goals and objectives should always be considered the first and foremost concern of all course planning, whether technology integration is involved or not, and change, course redesign, and resource development can be enhanced by technology tools but should not be driven by them. Effective teaching is a complex mixture of learning theory, the individual instructor's teaching philosophy and style, content expertise, the learning environment, the students, and the tools used for developing methodologies and strategies for teaching. Technology tools are just one of the possible tools that an instructor can use to help students construct knowledge and attain course objectives. Therefore the focus should be on what students will learn, how they will learn, and when they will learn.

Traditionally, in preparation for a career in higher education, faculty received little experience with, or exposure to learning theory and how students learn. Instead the traditional focus has been on their discipline specialty and how to present the specific discipline content. This practice however is slowly changing and today with all of the new technological advancements and changing perspectives, the lines between teaching practice, learning theory, and content delivery are becoming more blurred. New tools now available to faculty can facilitate a more interactive learning environment for both students and faculty. However, there are more details to attend to, more skills to be learned, and more options for delivering instruction. When students become active participants in the learning process, factors such as learning styles, motivation, and learning preferences also become more complex and important. As an instructional designer working with faculty in planning technology integration, I find that these more complex details and intricate interactions between the tools, student learning and the learning environment are the ones that can cause faculty the most problems.

Before thinking about the new options and tools and stressing over what needs to be learned in order to use the new tools, faculty should first focus on something with which they are very familiar, such as the course foundations and its goals and objectives. Taking the design and development of instructional materials one step at a time and designing a development plan with which they are comfortable, will help faculty better understand the process of redesign and they will have more success in the process of planning technology integration. As an instructional designer that works with faculty planning technology integration, the first step is to sit and talk. Not just about course goals and objectives but also about time frames for development, faculty technology skill levels and comfort levels as well as personal goals. Then when I have the total picture, I am better able to make recommendations to the faculty member about how to proceed, what will need to be done and how long the process will take. This process of mapping out a plan for course redesign and technology integration helps the faculty member to see the process in smaller increments so that it is not as overwhelming. Putting the redesign process into smaller chunks will also help faculty build confidence in their ability to deal with the new tools and technologies and facilitate a better success rate in the end.

2. Remember the Chickering and Gamson Good Practice Principles.

The findings of Chickering and Gamson (1987) that were published in the seminal article, " Seven Principles for Good Practice in Undergraduate Education" and have become a common framework or rubric for the design of, as well as, the evaluation and assessment of learning and teaching in higher education. These principles encourage student-faculty contact, collaboration among students, and active learning, as well as set standards for communication and expectations. Although the principles were not conceived originally for technology-related instruction, Chickering and Ehrmann (1996) contend that the "Seven Principles" can also be used as a guide for the design and implementation of teaching strategies utilizing new technologies. When using technology tools in the classroom, there is really no need to reinvent the wheel, or redesign a new way of teaching. Technology-based teaching and learning should be grounded in traditionally "good teaching" practices but yet take full advantage of the opportunities provided by the new technology tools. This process of melding the traditional teaching values and the new technology tools requires some creativity, flexibility and determination. Again, this comes back to the original goals and objectives of the course and how technology integration will enhance or detract from those original goals.

3. There is more to using technology tools in your teaching than learning software skills.

It is much easier to learn how to use the technology than it is to learn how to integrate the technology into classroom practice. Since the application of technology can vary from discipline to discipline and from instructor to instructor, there is not one right way to integrate technology tools, nor is there a tried and true method that will work for all situations. There are best practices and examples that can be customized for specific needs and disciplines, but there is not a process where a single procedure is mapped out and documented that will work in every case. Integrating technology into teaching practice requires experimentation, trying out new strategies and seeing which one is best for you. The nature of the task of experimenting and integrating the technology tools is also more risky for faculty since this is the stage where their experimentation becomes public and is tested in front of their class. Faculty can practice the skills in the privacy of their office, but eventually integration becomes a more public practice. This concept of learning something new and not being an expert at it immediately is a difficult concept for many faculty members to understand. Learning technology skills and how to integrate technology is a process that takes time. Faculty members are experts in their disciplines and can become frustrated when they do not have the same control here that they have in their content areas (King, 2003a). The process of technology integration also requires teamwork, respect, and trust as support staff, instructional designers and faculty work together to achieve course goals.

Learning how to create a PowerPoint, for example, is usually a linear step-by-step process that can be acquired easily with practice. Integrating technology, however, can be a non-linear process that involves many variables. It is not as clear-cut and straightforward as learning how to use a software application. Integrating technology also involves learning additional tasks such as how to use a video projector in the classroom and how to transfer your classroom files from one location to another. To learn how to create a PowerPoint, you work on your office computer. In order to present your PowerPoint in class there are other factors that interject more layers of difficulty that also will need to be mastered. Some questions that faculty may need to answer can be: How will you use the PowerPoint you created? Will you distribute the PowerPoint or an outline before class to students to help them prepare for the class, or will you make it available after the class for reference and review? Will the PowerPoint material be presented in a linear or non-linear format? When there is a familiarity with the software already, whether it be course management software or PowerPoint or any other technology tool, these other incidental tasks and decisions related to actually integrating the technology tools will not appear as monumental, so again it is helpful to break down the integration planning and learning into manageable chunks and skill sets that will not overwhelm or frustrate the faculty member.

4. Use technology tools to focus on and fix what is broken, not to tweak what is already working.

Rather than use technology to improve something that is already working in your class, I recommend using technology as a strategy to enhance something that is just not quite right, something that is not working. Use technology to help you solve a problem you could not solve otherwise. This approach will produce more dramatic results in the classroom and encourage continued use. In many cases faculty have a particular problem in their class, such as students not being prepared for class, or discussions in the F2F session are not active and stimulating. These examples would be good opportunities to try to use technology. For example, give short online pre-class quizzes to make sure students have read the material before coming to class and are prepared for discussion. You could provide an outline or PowerPoint slides for students to read, review and print out before coming to class.

You could have students email you their "muddiest point" concepts, or fuzzy points that they did not understand from the reading that could then be incorporated into the next class meeting for discussion (Brown, 2001). In many cases technology tools are available that can help facilitate more effective class management strategies and help you direct the learning so that the time spent in class is more efficient and effective. Technology can also be a tool that allows continued learning outside of the classroom by facilitating online discussion between the F2F class sessions. For example, if you have a stimulating discussion going in class and time runs out, set up an online discussion forum and encourage students to finish the conversation in an online format.

5. Start small first and pick low hanging fruit (start with easy to implement tools).

The best advice I can give to faculty just starting out with technology integration or just thinking about using technology is to start small. Many times faculty who are really excited and motivated about what they have learned about technology, tend to want to begin by designing complex projects or large projects. However, they often have trouble maintaining the successful implementation and momentum. It is best to start small, with just a lecture, or one activity in one topic and master that successfully before moving on to more comprehensive projects. An example of this scenario may be a faculty member who is interested in distance leaning. To put an entire course into a totally online modality is quite an undertaking and will require not just learning technology skills but also rethinking pedagogical issues related to teaching in cyberspace. If possible I would recommend that the faculty member first try an online discussion forum project in a traditional course, and test out the tools and environment before delving into a totally online course.

This suggestion to start small goes hand and hand with the fourth suggestion to fix what is broken first. You may have one component of your course that you have tried to rework but have had no luck redesigning - try technology here first. Focus on one part of the course, one that you can describe where the problem is, and why it does not work, and how you think it should work. Look for a component of a course to work on where it will be easy to see visible and measurable improvement. A successful experience with using technology and seeing a visible difference in student achievement or motivation will in turn motivate you to try something else. As you plan your implementation process, work on one area at a time, assess and then build on that success before you move onto something else. As an instructional designer I have witnessed more faculty success when planning and implementation are carried out in small chunks rather than in large comprehensive projects. As the comfort level with tools and skills increase faculty become more willing to take risks and continue to learn. One really bad technology integration experience can sour a faculty member's desire to try anything new again and therefore I try to monitor faculty successes and skill levels and step in to alleviate the frustrations and problems that may arise.

Another component related to this aspect of starting small is making sure that faculty know when, where and how to get the support that they need. Faculty should seek out support staff on campus to help them develop an implementation plan in stages that will help minimize risk and failure. Reliance on support people as a safety net and relationships built between the faculty member and support staff are critical components for technology integration success.

6. Build on and enhance skills you have before trying to learn all new software skills.

As confidence develops with technology tools and new technology skills are mastered faculty often think they will need to learn more and more software applications. In many instances it may be necessary to learn new software applications in order to achieve a particular course goal or objective, but that is not always the case. Software skills in one application can be built on and developed to a more advanced stage, which will also allow faculty to meet their course goals and objectives. For example, learning Macromedia Flash skills can help a faculty member develop instructional animations or multimedia products. Yet, you can also use more advanced PowerPoint features such as action settings, connecting presentations, using web links, and adding video and audio to achieve similar results. Using PowerPoint instead of learning a new program will empower the faculty member to "do it themselves" and take ownership of the projects. I have found more success when working with faculty to help them learn one software package better than to teach them only a little about several different software packages. Many times faculty think they must start completely over to learn a new application, but as a support person I try to help faculty understand that they already have skills that can be transferred and applied to a new software environment. For that reason, I try to teach faculty not just the step-by-step process of the software, but also help them understand how it works and why it works the way it does so that they can apply that knowledge to other similar software applications and situations. It is also helpful to use software applications from the same vendor that use similar tools, features and menus. For example, if you know Microsoft Word, there are many features of that application that are the same in PowerPoint or FrontPage. Once faculty realize that they know more than they thought they knew when trying out a new application they feel more comfortable and are more receptive to learning new features.

7. There is no ONE right way to use technology and no one killer application that will solve all your problems.

It would be great if we did have a "killer app" or "silver bullet" that would solve all our educational problems and enhance student learning, especially in this time of increased pressure on assessment and evaluation. However, there really is no miracle application or solution to technology integration. While one colleague may rave about a technological approach that has turned his classroom teaching around, it may not work in the same way in another discipline or classroom. Each technology tool and application may have potential for enhancing teaching and learning, but the real success is dependent on how the tool is integrated and if the use of the tool meets the original course goals and objectives. Remember, the best tool and the best integration plan are wasted time if they do not support the course goals and objectives. Many faculty members feel pressure to use technology tools in their teaching, yet technology tools may not always be the best option for the job. Careful consideration must be given as to why and how the tool should be used. The emphasis should be on the appropriateness of the situation and the tool, not just the tool itself. All of the factors discussed in this chapter such as students, teaching style, learning environment modality, student motivation, etc., are a part of the equation for a successful learning experience. There is no one factor that determines success and putting all your efforts into one component, such as the use of a technology tool, can lead to problems.

8. Using technology in your teaching requires a continuous commitment and updating of skills.

Integrating technology into your teaching is not a once and done deal. True technological literacy requires a continuous process of maintaining and updating skills and strategies. I would like to be able to tell you that putting in time to learn all the technology tools and skills at any one given point in time, will serve you for life. But this is not the case because inherent in technology is change. This evolving nature of technology can be a frustration for faculty who question why they should spend time learning something today that will change tomorrow. However, skills and strategies learned in one instance can in many cases be applied to other similar instances or new software applications. As an instructional designer, I try to help faculty find application for skills, and show them how to transfer skills from one situation to another. King (2003) states that technology training and integration can inspire lifelong learning attributes in faculty when they embrace what technology can do for them. A continuous commitment to technology literacy is necessary not only for learning new technology skills but also for maintaining the skills you have already learned. One of the biggest complaints that I receive from faculty trying to learn how to use to technology is they do not have enough time to continue to experiment and practice with the technology tools, and therefore they forget how to use the tools. I have found that faculty that appear the most comfortable with technology are also those that are motivated to keep learning, are continually using their skills and do not get rusty. This commitment on the part of faculty members to stay current with their technology skills should also be one that is recognized and supported by institutions and administration. Without support for their efforts, faculty will find it difficult to not only maintain their current technology skills but also to move ahead.

9. Think about the big picture (series of courses, high enrollment courses)–make it worth the effort.

In addition to thinking about a single course it is also helpful to think on a bigger and broader scale as you plan and update courses. What is the pre-requisite for your course? What strategies are used in that course that you can build on? What course follows your course and how can you integrate your content and strategies with what comes after? Using instructional design strategies and thinking about a series of courses, or multiple sections of a high enrollment course can facilitate a more impressive change effect. Collaborate and share the responsibilities and workload for integrating technology with other faculty. High enrollment courses with a large number of students are particularly good candidates for technology integration projects or course redesign. Spending considerable time redesigning or retooling an upper level course with a limited student enrollment is not as cost effective as one that involves more students, nor is it as rewarding an effort for faculty. High enrollment courses can also offer more opportunities for improvement. How can a large lecture course become more personalized with technology? How can group discussion and online projects create a stronger sense of community in a large lecture class? Technology tools can be used effectively for classroom management issues and communication that will in turn provide more quality and personal time in the classroom. Interdisciplinary courses are also good candidates for course redesign through instructional design planning and technology integration. Faculty collaboration across different disciplines can create an atmosphere more conducive for the development of unique and creative solutions for enhancing course curriculum.

10. Stay away from the flash of technology and focus on the functionality of the technology tool.

The last, but not least important guideline that I have to offer for planning and integrating technology tools into teaching and learning is to keep focused on the functionality of the technology tool. Chickering and Ehrmann (1996) state:

> Faculty members who already work with students in ways consistent with the Principles need to be tough-minded about the software- and technology- assisted interactions they create and buy into. They need to eschew materials that are simply didactic, and search instead for those that are interactive, problem oriented, relevant to real-world issues, and that evoke student motivation (para. 31).

As you become more visible as a technology adopter it is natural to want to become more visible in your use of technology and use more high-end technology tools. It is also understandable that experimentation with various tools and strategies may be warranted. But in order to remain successful with technology implementation, it all comes back to the main purpose of this article, which is to not lose sight of the original project or course goals and objectives and to use technology to support these goals and objectives.

References

Barr, R. B. & Tagg, J. (1995). From teaching to learning: A new paradigm for undergraduate education. *Change*, *27*(6), 12-26.

Brookfield, S. D. (1986). *Understanding and facilitating adult learning*. San Francisco, CA: Jossey-Bass.

Brown, D. G. (2000). *Teaching with technology*. Bolton, MA: Anker Publishing Company, Inc.

Brown, D. G. (2001). Faculty practice: The muddiest point. *Syllabus*, *November 2001*, p.22.

Brown, J. S., Collins, A. & Duguid, P. (1989). Situated cognition and the culture of learning. *Educational Researcher*, *18*(1), 32-41.

Chickering, A.W., & Ehrmann, S. C. (1996). Implementing the Seven Principles: Technology as lever. [Electronic Version]. *AAHEBulletin*, *49*(2), 3–6.

Chickering, A., & Gamson, Z. (1987). Seven principles for good practice in undergraduate education. *AAHE Bulletin*, *39*(7), 3-7.

Dale, E. (1969). *Audiovisual methods in teaching*. NY: Dryden Press

Dick, W. & Carey, L.M. (1978). *The systematic design of instruction*. New York: HarperCollins.

Fullan, M. (April 2000). The three stories of educational reform. [Electronic Version]. *Phi Delta Kappan (Online)*. Retrieved July 15, 2003 from http://www.pdkintl.org/kappan/kful0004.htm

Grabinger, R. S. (1996). Rich Environments for active learning. In D. H. Jonassen (Ed.), *Handbook of research for educational communications and technology*. (pp. 665-692). New York, NY: Simon & Schuster Macmillan.

Gros, B., et al. (1997). Instructional design and the authoring of multimedia and hypermedia systems: Does a marriage make sense? *Educational Technology*, *37*(1), 48-95.

Hannafin, M.J., Land, S.M., & Oliver, K. (1999). Open learning environments: Foundations, methods and models. In C. Reigeluth (Ed.),

Instructional-design theories and models: A new paradigm of instructional theory (Vol. II). (pp.115-140). Mahwah, NJ: Erlbaum.

Jonassen, D. (1991). *Computers in schools: Mindtools for critical thinking.* Pennsylvania State University Press.

Jonassen, D. (1991). Objectivism vs. constructivism: Do we need a new philosophical paradigm? *Educational Technology Research and Development, 39*, 5-14.

Jonassen, D. & Rohrer-Murphy, L. (1999). Activity theory as a framework for designing constructivist learning environments. *Educational Technology Research and Development, 47*(1), 61-79.

Kemp, J.E., Morrison, G.R., & Ross, S.M. (1997). *Designing Effective Instruction* (2nd ed.). Upper Saddle River, NJ: Prentice-Hall.

King, K. (2003). *Keeping pace with technology, Volume 2: The Challenge & Promise for Higher Education Faculty.* Hampton Press.

King, K. (2003a). Learning the new technologies: Strategies for success. In K. K. King and P. A. Lawler (Eds.), *New perspectives on designing and implementing professional development of teachers of adults,* (pp. 49-57), New Directions for Teaching and Learning, vol. 98. San Francisco: Jossey-Bass.

Knowles, M. (1980). *The modern practice of adult education.* Chicago, IL: Associated Press/Follett

Lankard, B. (1997). *New learning strategies for generation X.* Eric Digest. ED411414.

Lawler, P. A. (1991). *The keys of adult learning: Theory and practical strategies.* Philadelphia: Research for Better Schools.

Mezirow, J. (1991). *Transformative dimensions of adult learning.* San Francisco, CA: Jossey-Bass.

O'Banion, T. (1995). A learning college for the 21st century. *Community College Journal, 66*(3), 18-23.

Pascarella, E. & Terenzini, P. (1991). *How college affects students: Findings and insights from twenty years of research.* San Francisco: Jossey-Bass.

Reigeluth, C. M. (1983). Instructional design: What is it and why is it. In C. M. Reigeluth (Ed.), *Instructional-design theories and models.* Hillsdale, NJ: Erlbaum.

Reigeluth, C. (1996). What is the New Paradigm of Instructional Theory? *ITForum Paper 17*, Center for Online Education, University of

Georgia. Retrieved April 26, 2004 from http://it.coe.uga.edu/itforum/paper17/paper17.html

Sacks, P. (1998). No...Generation X is not OK. *The American Enterprise*, *9*(1), 46-49.

Schon, D. A. (1991) *The Reflective Turn: Case Studies In and On Educational Practice*, New York: Teachers Press, Columbia University.

Schuyler, G. (1997). *A paradigm shift from instruction to learning*. (Report No. EDO-JC-98-02). Office of Educational Research and Improvement (ED), Washington, DC. (ERIC document Reproduction Service No.ED414961).

Tapscott, D, (1998). *Growing up digital*. New York, NY: McGraw-Hill Book Company.

Toffler, A. (1980). *The third wave*. New York: Bantam Books.

Watts, M. M. (2003). Introduction: Technology as catalyst. In M. M. Watts (Ed.), *Technology: Taking the distance out of learning* (pp. 3-12), New Directions for Teaching and Learning, vol. 94. San Francisco: Jossey-Bass.

Wilson, B. G. (1997). Thoughts on theory in educational technology. *Educational Technology*, *37*(1), 22-27.

Winn, W. (1997). Advantages of a theory-based curriculum in instructional technology. *Educational Technology*, *37*(1), 34-41.

Integrating Technology into the Curriculum

Kelly L. Reimer
Elon University

Abstract

The purpose of this chapter is to explore the importance, challenges, and possible solutions and approaches of integrating technology into the higher education curriculum effectively and efficiently. Using current published literature, this paper provides an overview of the state of technology in higher education. Topics covered are the history of technology, learning theories, opportunities to redefine the educational model, challenges in the adoption of technology by faculty, concerns regarding student access to and skill level with technology, development of a strategic plan and implementation of that plan, and suggestions for future research.

Importance

Questions exist as to whether technology will have an impact on higher education. Skeptics and traditionalists believe higher education has been stable since the printing of the Gutenberg Bible in 1456. The technology of movable type, information storage, retrieval, and transmission through the written word has remained a constant, as well as the use of lectures and discussions (Oblinger & Verville, 1998). Prior to the technology of printing, Filippo di Strata declared in the fifteenth century, ". . . the world has got along perfectly well for six thousand years without printing, and has no need to change now" (Oblinger & Verville, 1998, p. 53). When radios, motion pictures, and television were invented, some believed these technologies would transform education. These technologies did have an effect on education, but none as great as the computer. Ten years ago, the computer was nothing more than a novelty item, but in today's world, the computer has been established as a relevant tool for education (Ferguson & Kopp, 1998).

Education literature forecasts the predictions of the value and power of technology. O'Banion (1997, p. 64) highlights some of these predictions:

- 'Technology is the primary vehicle by which institutions of higher education are going to re-engineer the teaching and learning process (Heterick, 1992, p. A17)'
- 'The eventual transformation of higher education and the integration of instructional technologies is inevitable (Gilbert, 1995, p. 47)'
- 'One can no longer afford to ignore technology and still maintain institutional health (Ehrmann, 1995, p. 24).'

Higher education is being asked to meet new needs. Our workforce needs to be more highly skilled, employment opportunities are growing rapidly in new technological industries, and older workers are returning to the university to update their skills (Bates, 2000). Duderstadt (1999) writes,

> Today's students are members of the 'digital generation.' They have spent their early lives surrounded by robust, visual, electronic media–*Sesame Street*, MTV, home computers, video games, cyberspace networks, ... and virtual reality. Unlike those of us who were raised in an era of passive, broadcast media such as radio and television, today's students expect–indeed, demand–interaction. They approach learning as a 'plug-and-play' experience: they are unaccustomed and unwilling to learn sequentially–to read the manual–and instead are inclined to plunge in and learn through participation and experimentation. Although this type of learning is far different from the sequential, pyramidal approach of the traditional college or university curriculum, it may be far more effective for this generation, particularly when provided through a media-rich environment (p. 7).

The learning needs of our students are changing dramatically, and their demand for technology at the higher education institution will become a competitive factor between institutions (Gilbert & Green, 1997).

At the same time, technology provides us with the tools to integrate learning and development theories into education. Ferguson and Kopp (1998) note that the research of education leaders, such as Jean Piaget, suggests that sophisticated learning can be accomplished by interweaving teaching and learning, having the learner participate in authentic learning environments, designing the learning environment to enable active student exploration and interaction, and allowing the learner to creatively apply new information. Computer technology today makes it easy to create authentic learning environments. Relevant computer-based activities are hands-on, and students can actively learn by conducting online research and developing multimedia presentations (Ferguson & Kopp, 1998).

Technology is providing the opportunity to redefine the educational model. Learning is no longer constricted by time or space, and the teacher no longer has the highest authority or a monopoly over knowledge (Hecht, Higgerson, Gmelch, & Tucker, 1999). New materials exist to facilitate the learning process, such as web-based instruction. George Watson is quoted as saying:

> ... these materials help take students out of the 'passive, listening-to-a-lecture' mode and prompting them to actively visualize problems from many perspectives. By interacting with a computer, with each other, and with instructors 'students develop skills with lasting value, such as effective communication, mastery of analytical thinking, teamwork, independent learning, and strategies for making the best use of all available resources' (Roselle & Pinholster, 1999, p. 17).

Redefining the educational model will improve pedagogical models. It will allow more tailoring to students with different backgrounds, learning styles, and prerequisite knowledge. It will also enable students to "study in an immense nonlinear space of knowledge" from all over the world (Larson & Strehle, 2001, p. 57).

Technology is a valuable learning tool, and research studies show that "it can make a measurable difference in student achievement, attitudes, and interaction with teachers and other students" (Oblinger & Verville, 1998, p. 117). Oblinger & Verville (1998) highlight some of the important research findings:

- Educational technology has demonstrated a significant positive effect on achievement. . . .
- Educational technology has been found to have positive effects on student attitudes toward learning and on students' self-concepts. . . .
- Introducing technology into the learning environment has been shown to make learning more student-centered, to encourage cooperative learning, and to stimulate increased teacher-student interaction.

- Courses for which networks were used increased student-student and student-teacher interaction, increased student-teacher interaction with lower performing students, and did not decrease the traditional forms of communication used (Oblinger & Verville, 1998, p. 117).

It's not the technology itself that is improving student learning, but rather that the students' needs are being better met through the use of technology. New educational environments and opportunities are now possible, and education is enhanced as a result (Larson & Strehle, 2001). If higher education institutions want to improve student learning through the use of technology, they must recognize the challenges in adequately preparing for this technology revolution.

Challenges

Today's higher education institutions are challenged with creating a new institutional environment to accommodate and promote the use of new technologies (Hagner & Schneebeck, 2001). Thirty-one and a half percent of all U.S. institutions indicated in a fall 2001 survey that integrating technology into instruction was the top IT issue confronting their campus (EDUCAUSE, 2002). At the same time, ". . . there is little, if any, public policy support for technological change for school districts, community colleges, or four-year institutions. The absence of policy support is felt in inadequate financial support for changes that technology might bring to schools" (Ferguson & Kopp, 1998, p. 31). Even when the technology may exist, there is no guarantee that it will become a part of the curriculum. Other challenges such as constant change, mismatched rates of change, faculty engagement, and mismatched resources confront today's higher education institutions (Gilbert & Green, 1997).

Heraclitus once said, "There is nothing permanent except change" (Stevenson, 1967, p. 232). What a true statement about technology! The speed at which technology changes is indeed very remarkable.

> Consider the automobile as an example of the transformative effects of technology. In 1985 the most expensive car made in the United States was a Cadillac. It cost $17,000, averaged 12 miles to the gallon, and weighed more than 1 ton. If the automobile industry had achieved the same technology trajectory as the computer industry, today a Cadillac would cost $12.63, weigh 14 pounds, get 5,900 miles to the gallon, and be 3 feet long! In fact, if you are

driving a Ford Taurus today, you are 'piloting' a vehicle that contains more computing power than the first lunar landing module (Oblinger & Verville, 1998, p. 43).

Technology is developing at an extremely rapid pace, and yet, consider this famous quote by an unknown author, "The pace of innovation in academe is perhaps best measured by the 25 years it took to get overhead projectors out of the bowling alley and into the classroom" (Gilbert & Green, 1997, p. 25). New applications of IT occur in monthly cycles, while higher education institutional structures change in a cycle based on decades or centuries, and variations in textbooks and publishing occur in years. This mismatch makes technology integration challenging (Gilbert & Green, 1997).

"When it comes to applying digital technology in the classroom, the biggest obstacle for higher education has nothing to do with resources or wiring. It's about faculty members" (Lynch, 2002, p. B15) and their level of engagement. Faculty members have always had tremendous autonomy within the higher education institution. With a variety of motivations, their approaches to tasks are quite different. As the tenure process draws to a close for a faculty member, the influence of administration on them decreases dramatically (Hagner & Schneebeck, 2001).

Based on Everett Rogers' theory on the diffusion of innovation, Hagner and Schneebeck (2001) offer four classifications or waves that can be useful in understanding faculty and their adoption of emerging technologies. Although these classifications are simplistic, and some faculty possess characteristics of multiple groups, most faculty demonstrate predominant characteristics of one of the four groups. The first wave of faculty members, are the 'entrepreneurs.' These faculty members are at the forefront of adopting new technology and taking risks to use it in the classroom, not just for the sake of using it, but to follow through with their own commitment to quality teaching and learning. They are not seeking rewards for their ventures, but can become discouraged when their work is not recognized. The second wave of faculty members, are the 'risk aversives.' Sharing a commitment to the quality of teaching and learning with the first wave, their actions are not as profound because of a lack of technical expertise, fear of losing their current success in teaching with the move to using technology, need for instructional support, and hesitancy for the process of self-examination by development of new learning objectives. The third wave of faculty, are the 'reward seekers.'

These faculty members exist at institutions where the tenure and promotion structures reward the integration of technology into the classroom. Faculty wanting to advance in their careers will use the technology to advance their teaching specifically for self-improvement. The last group of faculty, are the 'reluctants,' those faculty who are computer illiterate or "believe the traditional models of learning are superior" (Hagner & Schneebeck, 2001, p. 2-4). Carey and Dorn define the 'reluctants' as those people with 'technophobia.' Their fear of technology exists because "the technology does not meet three criteria of acceptance: familiarity, usefulness, and control" (Carey & Dorn, 1998, p. 71). The only way to overcome technophobia is complete immersion, until the technology becomes "familiar, useful, and controllable" (Carey & Dorn, 1998, p. 71).

Specializing and having extreme competency in a specific academic area, faculty members tend to feel uncomfortable, vulnerable, disempowered, and frustrated when they confront new technologies. New technologies are oftentimes not easy to use and are unreliable. Impatient or uncompassionate technology support can oftentimes lead faculty to eliminate any possibility of using technology in their classroom (Gilbert & Green, 1997). Technology is a source of stress for the majority of faculty members, and so they just do not use it. Students know more about technology than faculty members do in most cases, and it is "downright unthinkable" that students hold this power over faculty (Lynch, 2002, p. B15). The Campus Computing Project conducts nationwide research on information technology in U.S. Higher Education (EDUCAUSE, 2002). One of their recent surveys indicated that in more than one-third of the courses, faculty still do not use e-mail, much less integrate other forms of technology into the classroom (Lynch, 2002).

At the same time, technology changes the teaching and learning process from teaching-centered to learner-centered (O'Banion, 1997). As technology is more deeply integrated into the classroom the role of faculty members changes from teachers who transmit intellectual content by lecture format to consultants/coaches who design learning experiences, processes, and environments to inspire, motivate, and manage active learning processes (Duderstadt, 1999). Some faculty members do not want to give up their "gate keeping function" (Donald, 1997, p. 226). Those faculty members who are willing to make some changes lack the necessary skills and are approaching technology integration by using technology to teach specific topics and introducing e-mail into a course. This type of integration does not come close to the potential that exists, but rather just shifts the current form of instruction already used in the classroom to the computer. Technology is being used in old ways (Gillespie, 1998).

Surprisingly enough, there is also a remarkable variation of the computer skills students possess upon entering college (Gilbert & Green, 1997). Such variation in skills can be seen by examining the following facts:

- "The Committee for Economic Development reported in 1995 that more than two-thirds of American families with incomes over $50,000 and college-educated parents have a computer at home" (O'Banion, 1997, p. 70). At the same time, even though "access continues to grow, it is strongly related to income, gender, and profession. Many of those who do have computers have machines that are not suitable for multimedia or Internet access" (Bates, 2000, p. 86).
- "Socioeconomic patterns of computer ownership are significantly different from those for other domestic audiovisual equipment. . . . Low-income earners have traditionally been the early adopters of color televisions, laser discs, audiocassette players, compact disc players, and so on. This is not true, though, of personal computers; low-income families have preferred to buy video and computer games instead" (Bates, 2000, p. 86).
- At public U.S. universities in fall 2001, 65.4% of students had or owned a desktop computer, and 19.4% of students had or owned a laptop computer. At public 2-year institutions, computer ownership is significantly lower (EDUCAUSE, 2002). Between institutions and within institutions, these percentages are likely to vary. Institutions who serve disadvantaged groups will most likely find that their target groups have less access to computers compared to an institution with high tuition fees and difficult admission requirements (Bates, 2000).
- In the United States, "the majority of entering college freshmen reported a half-year or more of some kind of computer training in high school" (O'Banion, 1997, p. 71). Yet, "many schools still do not have adequate computer provision, so students are entering the university or college with widely different levels of com-

> puter literacy . . . ownership of a computer does not necessarily mean the student has the necessary computer skills for studying" (Bates, 2000, p. 87).

Institutions must be careful about assumptions they make regarding the technology competency levels of their students. They must also be aware of the ownership of computers by their student body. The issue of ownership is another chicken-and-egg issue. "If students are not required to have a computer for their studies, they are less likely to purchase one. If students who are thinking of purchasing . . . are not given clear specifications as to what is needed, they are more likely to purchase a less powerful model" (Bates, 2000, p. 87).

Solutions and Approaches

In a 1998 U.S. national survey it was found that only half of higher education institutions in the United States have a strategic plan for technology, and most of those plans only discuss the infrastructure (networks and hardware) at an institutional level. Only 40% of institutions had some sort of instructional plan for integrating technology into the curriculum. Strategic planning is usually prepared at the institutional level, but when it comes to integrating technology into the curriculum, the approach must occur at all levels, and the plans of the separate areas must fit together. Not only that, but everyone at the institution must "buy-in" to the plan (Bates, 2000). The development of such a plan will require debate, time, and leadership (Daniel, 1997).

Where should all of this planning begin? The best place to start is at the departmental level by asking each department chair to develop with their faculty a three- to five-year teaching plan that covers all forms of instruction, including technology-based education (Bates, 2000). The department chair must lead this process by inspiring a vision that is attractive to all concerned (Daniel, 1997). Departments cannot escape addressing the issue of the use of new technologies, as personal computers and the Internet are a part of everyday life. Department chairs can play an extraordinary role in leading their department to develop a strategic plan.

> No department would–nor should it–attempt to utilize all the pedagogical alternatives ... What departments do need to do is to look at their curricula in combination with the options in the pedagogi-

> cal smorgasbord to choose the means most appropriate to its discipline. Development will remain haphazard unless you [department chair] help your department focus on this matter as something of importance (Hecht et al., 1999, p. 14).

The department chair serves as the "link between faculty and staff, and between policy and implementation" (Hecht et al., 1999, p. 49). They must help their faculty to "work collaboratively with colleagues as well as with technicians" (Hecht et al., 1999, p. 149).

The faculty must support these technology plans for them to work. Ideally, a plan needs to be developed that will inspire strong commitment to a high-quality strategy (Daniel, 1997). Faculty engagement is one of the largest problems in the integration of technology into the curriculum. Faculty will fear that these technology plans will pull them away from research, are just a way for administration to reduce expenditures or resources, and are another way to add additional work to the faculty (Bates, 2000). Some faculty will even question if new technologies will put them out of work (Ashby, 1974). Technology plans are likely to affect every member of the institutional community, and concerns can be eased by including everyone–departmental meetings, subcommittees, task forces–and at every level–administrative, technical, secretarial staff, and students in the planning process. The preparation is "probably more important than the actual details of the plan. As General Eisenhower said, 'Planning is everything; the plan is nothing'" (Bates, 2000, p. 48). If everyone is aware of the upcoming changes, fundamental modifications in teaching can occur, rather than just "drifting into technology for teaching" which can "be dangerously expensive and ineffective policy" (Bates, 2000, p. 48).

In developing and implementing the plan these things must be considered:

- The environment–What are the external and internal environments each department is likely to encounter (finances, enrollments, trends, activities of competitors, current strengths and weaknesses, opportunities and threats)?
- Developing a vision for each department–Get faculty to think outside the box by providing half-day workshops demonstrating innovative teaching and/or presentations or demonstrations by outside speakers.

- Developing a vision for the institution–The senior management of the institution must have "provided an institutional vision or context for change within which the department's vision can 'nest'" (Bates, 2000, p. 50). The vision must include an implementation plan to ensure that concrete actions will take place.
- Ensuring implementation is occurring–Once the plan has been decided on, the plan must be put into print and made accessible to all parties involved. The departmental chair or the vice president at the institutional level must verify on a regular basis that the plan is developing and that appropriate adjustments are being made if necessary (Bates, 2000).

At the institutional administrative level, institutions must do several things to ensure that the plan can be implemented (Bates, 2000). First, institutions must examine what they already have access to, determine the competencies of their employees and students, and provide opportunities for development (Daniel, 1997). The appropriate support staff must be made available, such as media production specialist, instructional designers, delivery and student support, multimedia developers, etc. Faculty and students must have the necessary access to technology. Faculty must be trained and their skills need to be developed in the use of technology. Studies find that institutions should not depend on faculty development as the main strategy for technological change; instead, faculty must be immersed in an environment of teaching and learning with technology. At Virginia Polytechnic University, 96% of all faculty members participate in a faculty development program, and 50% of all faculty members had attended a second faculty development program until 1999. Administration encouraged faculty to attend the development program by linking the acquirement of a computer to participation in the program. Recognizing that different faculty members are at different places in their use of technology–phobic to expert– strategies must be developed to address and meet the needs of each faculty member. Strategies may consist of policy (tenure concerns), providing comprehensive and systematic technical and professional support, and rewarding creators of technology-based teaching and learning materials (Bates, 2000). The administration must provide release time to faculty who are developing their curriculum for the new teaching technologies (Daniel, 1997). Faculty members need to be taught how to evaluate the effectiveness of different teaching technologies. Bates (2000) uses the acronym ACTIONS to help faculty in this process:

- "*A*ccess and flexibility: How accessible is a particular technology for learners? How flexible is it for a particular target group?
- *C*osts: What is the cost structure of each technology? What is the unit cost per learner? How do costs differ between technologies within a particular context?
- *T*eaching and learning: What kinds of learning are needed? What instructional approaches will best meet these needs? What are the best technologies for supporting this teaching and learning?
- *I*nteractivity and user-friendliness: What kind of student interaction does this technology enable? How easy is the technology to use?
- *O*rganizational issues: What are the organizational requirements, and the barriers to be removed, before this technology can be used successfully? What changes in organization need to be made?
- *N*ovelty: How new is this new technology? How reliable is it? How will this technology contribute to institutional renewal?
- *S*peed: How quickly can courses be mounted with this technology? How quickly can materials be changed?" (p. 201).

As faculty begin to implement technology into the curriculum, the core technology competencies for students can be introduced and expanded upon in almost every course. Students who may lag behind in their technology skills will learn how to use the computer effectively and efficiently. As students gain these skills, they will come to appreciate the asynchronous nature of the medium and will find "technology both liberating and more personal than traditional instruction" (Oblinger & Verville, 1998, p. 104).

Summary

The purpose of this chapter was to explore the importance, challenges, and possible solutions and approaches to integrating technology into the higher education curriculum effectively and efficiently. Technology is a part of our everyday lives, and upon graduation our students must have the necessary skills and competitive advantage to work and live successfully. Integrating technology into the curriculum is the only way to ensure that our students leave higher education institutions with computer literacy, computer competency, and information literacy (Oblinger & Verville, 1998). Even though technology is important to our students, many higher education institutions lag behind in teaching these skills to their faculty, staff, and students. Many challenges exist as to how to integrate technology into the curriculum and to get the buy-in of faculty, staff, and students (Hagner & Schneebeck, 2001). Institutions can take lots of different approaches to meet these challenges with solutions. Strategic planning at all levels of the institution is crucial to the successful development of the vision and the implementation of the plan (Bates, 2000).

Conclusions

Several higher education institutions have experienced successful implementation of technology into the curriculum. These schools, such as the University of Delaware (Roselle & Pinholster, 1999), University of Michigan (Lynch, 2002), University of Texas at Arlington ("School Requires," 1999), Old Dominion University, and Florida State University (Mendels, 1999), have developed computer literacy requirements of their students and have integrated technology into the curriculum throughout a student's four years of study. Accrediting agencies, such as the Commission on Colleges of the Southern Association of Colleges and Schools, have encouraged institutions to develop policies regarding the use of technology (School, 1999) and have added the words "the basic use of computers" to the list of skills graduates of institutions must demonstrate (Mendels, 1999).

To help higher education institutions integrate technology into the curriculum, there are several key things that need to be facilitated:

- K-12 education needs to include more technology in the classroom to prepare students coming into college. In a 1996 survey, only half of United States teachers had ever used a computer at all (Flowers, 2000).
- Most of the research that has been conducted on information literacy is a part of library literature, "which is seldom read by educators or state decision makers" (Breivik, 1998, p. 132).

More research needs to be conducted on the following:

- The value added to having a computer literacy/competency

- Determining the impact of resource-based learning within higher education on a student's academic success
- Information needs of different community groups
- Impact of information literacy on work force productivity and the quality of people's lives
- How information can be more effectively managed to meet educational and societal objectives
- How information management skills impact higher education institutions' overall school and academic performance (Breivik, 1998).

At this time, higher education institutions will have to take risks to integrate technology into the curriculum. Institutions will need to recognize their current resources and utilize them as best as possible. These resources include faculty members who are entrepreneurs, students who excel with the use of technology, or the support of instructional designers who can work one-on-one with faculty. Students have an "800 number, ATM mentality: 'When students can get cash at 2 A.M., download library materials at 3 A.M., and order shoes from L.L. Bean at 4 A.M., it is only educational inertia that keeps them convinced that they must learn calculus by sitting in the same classroom for fifty minutes, three times a week'" (Blustain, Goldstein, & Lozier, 1999, p. 54). The integration of technology should occur because, as Dianne Lynch, a faculty member, stated, "We should do it because there is nobody better prepared or better qualified to teach–now and from now on–and nobody more dedicated to maintaining the rigor and quality of student learning" (2002, p. B15). By taking risks, higher education institutions will find ways to integrate technology into the curriculum, and ultimately, to enhance the academic experiences of faculty and students.

References

Ashby, E. (1974). *Adapting universities to a technological society.* San Francisco: Jossey-Bass Publishers.

Bates, A.W. (2000). *Managing technological change: Strategies for college and university leaders*. San Francisco: Jossey-Bass Publishers.

Blustain, H. Goldstein, P., & Lozier, G. (1999). Assessing the new competitive landscape. In Katz, R. and Associates. *Dancing with the devil: Information technology and the new competition in higher education*. San Francisco, Jossey-Bass, Inc.

Breivik, P.S. (1998). *Student learning in the information age*. Phoenix: American Council on Education Oryx Press.

Carey, K.T. & Dorn, S.M. (1998). Overcoming obstacles through use: A case study. In Gillespie, K.H. (Ed.). *The impact of technology on faculty development, life, and work*. (pp. 67-78). San Francisco: Jossey-Bass Publishers.

Daniel, J.S. (1997). *Mega universities and knowledge media: Technology strategies for higher education*. London: Kogan Page.

Donald, J. (1997). *Improving the environment for learning: Academic leaders talk about what works*. San Francisco: Jossey Bass Publishers.

Duderstadt, J. (1999). Can colleges and universities survive in the information age? In Katz, R. & Associates. *Dancing with the devil: Information technology and the new competition in higher education* (pp. 1-25). San Francisco: Jossey-Bass, Inc.

EDUCAUSE. (2002). *The pocket guide to U.S. higher education: 2002* (October ed.) [Brochure]. Washington, D.C.: EDUCAUSE.

Ferguson, K.M. & Kopp, O.W. (1998) *Technological collaborations: K-12 and higher education*. Bloomington: Phi Delta Kappa Educational Foundation.

Flowers, C.P. & Alogozzine, R.F. (2000). Development and validation of scores on the basic technology competencies for educators. *Educational Psychological Measurement*, *60*, 411-418.

Gilbert, S.W. & Green, K.C. (1997). Moving information technology into the classroom. In Enhagen, L. (Ed.). *Technology and higher education* (pp. 25-56). Washington, D.C.: National Education Association of the United States.

Gillespie, F. (1998). Instructional design for the new technologies. In Gillespie, K.H. (Ed.). *The impact of technology on faculty development, life, and work* (pp. 39-52). San Francisco: Jossey-Bass Publishers.

Hagner, P.R. & Schneebeck, C.A. (2001). Engaging the faculty. In Barone, C.A. & Hagner P.R. (Eds.) (2001). *Technology-enhanced teaching and learning: Leading and supporting the transformations on your campus* (pp.1-12). San Francisco: Jossey-Bass Publishers.

Hecht, I.W.D., Higgerson, M.L., Gmelch,W.D., & Tucker, A. (1999). *The department chair as academic leader*. Phoenix: American Council on Education Oryx Press.

Larson, R.C. & Strehle, G.P. (2001). Edu-Tech: What's a president to do? In Goodman, P.S. (Ed.). *Technology enhanced learning: Opportunities for change* (pp. 21-59). Mahwah: Lawrence Erlbaum Associates.

Lynch, D. (2002, January 18). *Professors should embrace technology in courses*. The chronicle of higher education, B15. Retrieved August 27, 2002, from http://chronicle.com/cgi2-bin/printable.cgi.

Mendels, P. (1999, September 29). *Universities adopt computer literacy requirements*. New York Times. Retrieved November 3, 2002, from http://www.nytimes.com.

O'Banion, T. (1997). *A learning college for the 21st century*. Phoenix: American Council on Education Oryx Press.

Oblinger, D.G. & Verville, A. (1998). *What business wants from higher education*. Phoenix: American Council on Education Oryx Press.

Roselle, D.P. & Pinholster, G. (1999). Capturing the first keystrokes: A technological transformation at the university of Delaware. In Oblinger, D.G. & Katz, R.N. (Eds.). *Renewing administration: Preparing colleges and universities for the 21st century* (pp. 3-19). Bolton: Anker Publishing Company, Inc.

School requires computer literacy. (1999, August 17). USA today. Retrieved November 3, 2002, from http://www.usatodaycom.

Stevenson, B. (Ed.). (1967). *The home book of quotations: Classical and modern*. (10th edition). New York: Dodd, Mead & Company.

From Talking At You to Talking With You: Reshaping PowerPoint for Interactive Learning

Ralph Olliges
Webster University

Sebastian Mahfood
Kenrick-Glennon Seminary

Roy Tamashiro
Webster University

Abstract

Technology-mediated transactional learning practices make it possible for learners to intimately engage their course materials, their instructor, and their classmates across time and space. Although PowerPoint may be ensconced in stereotypical applications in this age of cyber-frenzy, creative instructors have used PowerPoint and other tools to renovate the electronic teaching-learning environment, whether it is the face-to-face or online classroom. By seeing PowerPoint as a malleable tool, instructors can increase dialogue within the classroom and transform the teaching process from a transference- and transmission-based instructional model to one that is truly transactional.

Introduction

Today, we see an increasing number of higher education faculty members using media-rich materials in their teaching and learning environments (Olliges and Mahfood, 2003). The primary reason for this is that colleges and universities are entering into collaborative relationships with content deliverers to develop distributive learning networks that can meet the needs of an increasingly diverse community of learners. Instructional technologists are being hired, consequently, to strengthen the development of computer-mediated teaching and learning environments. The trend, therefore, is moving in the direction of the institutional expectation that the professorate will engage in training on the new tools available for its use. Moreover, students who have grown up in a multimedia age of content delivery have raised their expectations and practically demand that professors employ multimedia in their classes (Rickman and Grudzinski, 2000) and provide them with various means with which to interact with the course materials. Catering to these needs are software companies like Microsoft and Corel that make slideshow programs like PowerPoint and Presentations[1] user-friendly and cheaply available through academic pricing.

The increasing use of multimedia in college classrooms has begun to have an effect beyond merely facilitating lecture materials; it has also led many instructors to perceive the need to change their teaching methods from the traditional didactic, lecture-oriented format to dialectical, discussion-rich interactive approaches (Hoover, 1980; Newble and Cannon, 1995; and Orlich, et. al., 2004). Even in large group settings, instructors seek creative ways to engage students in the material and get them to actively participate. In the past, this was done in a variety of ways—by breaking average-size classes into small discussion groups or by dividing lecture-hall-size classes into labs. Students have been asked to package parts of the lecture materials for presentation to the rest of the class in the hopes that they would develop a facility with the material if requested to present it to other students. Instructors have mostly rejected assessment practices based on rote-memorization techniques and engaged students in collaborative and project-based learning opportunities to teach them how to learn beyond the time and space allotted for the class (Harris et alia, 2003). Regardless of the methods used, it is helpful to engage the students in their own learning processes in order to create a learning environment that supports students with various learning styles and interests.

The idea, then, is that the use of presentation software should enable professors to accomplish both tasks at the same time—to bring different media into the classroom environment and to create an interactive teaching and learning environment. Presently, though, both students and faculty are dissatisfied with the often *disengaging* effects of using PowerPoint for this kind of teaching (Quibble, 2002). The reality is that this has nothing to do with the tool being used for content delivery but with the fact that the tool is being used to replicate, not transform, the traditional lecture method. Using PowerPoint to create bullet points of the lecture so that students may copy them into their notes does not transform the teaching and learning environment. What is required is a truly transformative pedagogy. The software will not magically provide that to a teacher who does not continue his or her own education on building teaching and learning environments for students. This article aims to provide some ideas for college-level educators on how non-stereotypical uses of PowerPoint may assist them in making their instruction more interactive and dialogue-rich.

From Didactic to Interactive Instruction

Typically, instructors learn to use PowerPoint in a few hours as they convert lecture outlines into a series of animated bullet points, supplemented with clip art, photos and/or sound effects. The instructional paradigm of this approach may do nothing beyond replicating the lecture since the sequential point-by-point explanation of course materials is the most natural way to convey information under a transmission theory of communication. This is comparable to a professor transferring course notes onto transparencies to be viewed with overhead projectors. This approach saves the instructor the in-class time of writing his or her notes on the classroom boards. While some teachers were able to use overhead projectors to engage student activity and response, most used overhead projectors to convey information. PowerPoint is typically used for the same purpose.

PowerPoint itself is not transformative. There were lots of previous methods of content display, and PowerPoint can be used to replicate the lecture materials just like any of the former methods. While PowerPoint has features that overhead projectors and chalkboards may not, such as color graphics, photographs, video, audio, and animation, these features do not in themselves make the presentation interactive. Instead a transformative pedagogy requires the instructor to (1) design the lesson to include student involvement and dialogue in the presentation, and (2) find creative ways to shape the PowerPoint sequence so it becomes something other than the usual linear, didactic presentation. The transformation, then, has to come in using PowerPoint to move beyond the lecture to make the instructional experience interactive. Since using PowerPoint does not automatically give instructors interactive experience, they must change their teaching methods along with the changes they make in the tools they use.

Ultimately, the professor aims to change from the transmission model of content delivery to the transaction model of student interaction with the content. Dr. H, for instance, teaches a course on the history of science. In watching the public television NOVA program "Secrets of Photo 51," she was inspired to have students examine the claim that the molecular biologist Rosalind Franklin deserved more credit than she gets for her research into the DNA molecule. The idea of making an interactive slide of the now famous x-ray photo of the DNA molecule "photo 51" came to her after she learned that James Watson and Francis Crick had won the Nobel Prize upon seeing Dr. Franklin's data, including the key "photo 51," without her permission. According to this claim, Franklin's data led Watson and Crick to the discovery of the double helix. Dr. H thought that integrating the Internet with PowerPoint would stimulate class discussion. She placed transparent hotspots over various parts of the photograph and hyper linked them to enlargements of the sections or to other documents including URLs that gave more details or explanations about the x-ray photo.

In class, Dr. H projected the PowerPoint slide and used the hyperlinks to explain the photograph and to invite student questions and dialogue about the phenomenon. This procedure could also be done using transparency overlays on an overhead projector, but what PowerPoint provides is the hyperlink to content that would later be readily accessible on a student's own computer.

The new idea, then, is to use these links from a single image to discuss and explore the topic in class as a bridge to assisting further research outside of class. Students appreciate this method because it provides them with an authentic way to respond during the dynamic of an instructor-led presentation, for it is they who direct the instructor and learn to use the teacher as an agent in their learning process rather than their being used by the teacher as audiences for canned lecture materials.

Teachers as well as students learn best by involving themselves with the materials being studied and with the process by which those materials are presented to them in a classroom setting. Our example can be called a transactional educational model because it relies on interaction within a community as opposed to the transmissional educational model that relies on content delivery from an individual to an audience. The medium being used is not one that delivers content through a sequential and systematic method but one that negotiates that content within a community of learners. The medium, then, is not the technology but the environment that technology helps to create. By asserting that we can shape PowerPoint to not be didactic and sequential, but to be interactive and nonsequential, we are not changing Marshall McLuhan's 1964 dictum that the medium is the message. What we are changing is the focus from the technology to the people. The technology merely enables the audience to interact more fully as the audience explores its own interpretations of the material. Certainly, the instructor's intention to focus the audience on a particular area of exploration remains through the instructor's arrangement of the course materials. The links chosen for exploration also convey the rhetorical strategy of the instructor. In these ways, the audience remains within the paradigm constructed by the instructor, but the fact that the audience expresses its voice enables it to contribute to the shaping of the discourse. The members of that audience can assert their own rhetorical ends, and it is here that learning as a negotiated reality takes place.

The medium, then, is not so much governed by PowerPoint as it is informed by PowerPoint which is just a tool used to help create it. The channel has shifted from the tool to the classroom discourse itself—to the ability of the student to respond to it in the same language used by the medium. The medium, as expanded to include the interactivity of the audience, the instructor and the PowerPoint, continues to be the message. This triune medium that makes up the transactional model involves us interactively with the object we are studying, which is why it captures our interest more than the transmission model does. In the past, we were limited to being consumers of the technologies that surrounded us. In the present, we can be producers of those technologies in our interactions with others.

The sequential strategy of the transmission model relies on the intentional communication of the instructor to build a lecture from one point of explanation to the next. PowerPoint is an effective tool to replicate this method because its slides also move from one point of explanation to the next as the instructor proceeds along the path of the lecture. Hence, PowerPoint facilitates the process of student discovery. Any given

PowerPoint slide can have options in the form of links from which students might select new directions for the lecture—each of these links leading to new PowerPoint presentations that also include links beyond them. This would be analogous to a choose-your-own lecture format where various areas of exploration lead to various ends of discovery—which works as long as the instructor is agreeable to guiding the class in any of the directions the students wish to pursue.

Because this method requires the preparation of a lot of miniature PowerPoint shows all strung together, some professors like Dr. K, who teaches classes on prophetic literature and the Pentateuch, have preferred to embed the technique of inquiry into a single presentation template to save preparation time. Before she learned about presentation technologies, Dr. K had lectured to the students, had asked questions of them in class concerning their readings and had required a group project of them due at the semester's end. Over time, Dr. K noticed that her lecture-based "knowledge-transmission" format resulted in the students' not internalizing the course materials. Instead, they were going through the motions of what was expected of them and memorizing the areas on which she had placed the most importance. When Dr. K tried putting her lectures on PowerPoint, the structure continued to be a sequential point-by-point lecture. Her question became how to use the tool to achieve greater student participation and interaction with the course materials. She answered this by interspersing throughout the slide sequences a new screen with a question typed in a large font, such as "What is the book that precedes Ezekiel in the Bible?" The students would open their bibles and find that some bibles had different books preceding Ezekiel. A half dozen PowerPoint screens then followed that engaged student discussion through the results they had discovered. Inserting these questions broke the monotony of the lecture and reminded Dr. K. to include student discussion throughout. The transmission format would have used one screen to list the reasons for the differences and would have then advanced without ceremony to the instructor's interpretation of what that means. In effect, the transmission format would have conveyed more information in a given hour of class time, but it would not have prepared students to handle the information given them. Students who have time to absorb the materials provided during the time they are learning them found that they not only retained them better but also better understood the concrete applications for them in the area they were studying.

Because instructors perceive that they must convey all the information they have within one semester, many resist moving away from a transmissional methodology. Lecturers tend to like PowerPoint because it naturally facilitates the lecture environment. The real struggle is to convince lecturers to give up lecturing to their students and to start engaging their students as producers of class dialogue. It is difficult for instructors like Dr. K to break familiar and comfortable habits of teaching to include unfamiliar techniques. Using PowerPoint to change one's teaching merely from the straight lecture format to a lecture-discussion format illustrates Walter J. Ong's view in *Orality and Literacy* (1982) of the transformative nature of technologies (PowerPoint, in this case) on the human consciousness. Working with the PowerPoint application can actually assist professors who are usually not adept in educational theory discover and develop instructional strategies like intentional questioning, and through these strategies move from a transmission-based to a transaction-based pedagogy.

Since PowerPoint is such a comfortable program for instructors to learn and use, it functions as a transitional object and its use, in effect, is the doorway into not only the use of technology in the classroom, but also into interactive, dialogic instructional strategies. Programs like PowerPoint work on the principle of graduated complexity. The more an instructor wants to do with the program, the more an instructor will be able to do with the program. The instructor does not panic to master the program before its first use. This leaves open a wide learning curve to enable the instructor to learn the program through its continued use. Prof. W teaches a Holocaust unit in his history course in which he had become accustomed to using a slide projector to show some 80 photographs of Nazi genocide practices. Prof. W would traditionally interpret each slide, often writing names and dates on the board. In the PowerPoint version of this presentation, however, the keywords were typed onto or next to the images he displayed. Instead of spending time writing the words on the board or an overhead transparency, he could maintain face-to-face contact with the class and spend more time asking engaging questions or facilitating inspiring discussions. To do all this, Prof. W did not have to learn everything about PowerPoint. He only had to learn how to cut and paste a scanned image into the PowerPoint slide and how to type his text next to it. When Prof. W is ready to follow Dr. H and embed hyperlinks within his images, he will then have grown beyond his previous instruction. Even without the hyperlinks, Prof. W has improved his lesson significantly, and it took only half an hour for him to learn the software. His students, he says, actually come to class referencing the audio materials he posts for the library reserve versions of his PowerPoints, and their conversations over things like the French suppression of the Knights Templar carry over into the lunchroom, which, he thinks, is an exciting development for any teacher.

Instructors who have been lecturing for many years often find themselves comfortable with the idea of breaking their classes into discussion groups and investing class time in interactive, group-based presentations. Fr. H, for instance, who teaches a course in Pauline literature, used to read his lectures to his students without looking up from his notes to engage his audience. After learning how to use a presentation software program, Fr. H realized that he could get the students to interact with projected text and images in ways they had never interacted with his 75-minute lectures. He realized that students had good ideas to contribute to the development of the course, so he had students do class presentations using PowerPoint and other presentation programs. It was his first learning what these presentation techniques would mean for him that brought about a revolution in his teaching. He overcame the barrier of his not having had strong presentation skills and had what other members of his department have come to call "a Pauline conversion" in teaching methodologies. The main change can be seen in how his students no longer simply react to his teaching but actively engage it through the transactional environment he established when he invited student participation and leadership during class time. Instead of the students being consumers of the New Testament scripture he teaches, they are actively involved as producers of interpretive discussions that have Fr. H as their guide and mentor.

A true dialogical style develops when the instructor consciously allows the dialogue with the students to shape and transform the lecture. This transformation of an instructor's method of delivery from the monologic to the dialogic can occur, of course, without the use of PowerPoint. The method does not need any one kind of technology for it to work. However, presentation through PowerPoint starts the transformative process in the sense that it distributes the focus of the students to both the instructor and the projection screen. The extension of the instructor to a spot six feet away from where he or she is standing refocuses the emphasis of the lecture every time the instructor clicks his or her mouse. It is not that the instructor becomes de-centered. Instead, the materials the students are being asked to engage become central to them and the thing to which they respond. The discussion, then, released from the central authority figure, more easily flows over class time.

This interaction between the instructor and the students through PowerPoint occurs when PowerPoint is used in anticipation of student dialogue. The delivery of a given concept, then, begins with the title of the concept, some information concerning it and dialogue space whereby the instructor has some options for discussion. In one case, the instructor might call upon the students to do critical, deductive thinking that leads them from the general concepts of the topic to specific instances of its applications or value. If the students are being lectured about a historical concept, for instance, the slide title can point to a concept and the next three clicks can bring up specific details of that concept. Before the details on which the instructor would like to focus are revealed, the students can be asked for their thoughts on what those details might be with the instructor coaching them through dialogue in the direction he or she is heading. Through this process, some details might arise that the instructor had not listed, and that brings them to the table for discussion. As the instructor reveals the details, the students are involved in processing what they have learned along the boundaries the instructor has imposed. Their discovering these boundaries through dialogue with the instructor makes the material contained within those boundaries more meaningful to them than it would have been had the instructor just enumerated three or four points for inclusion.

PowerPoint enables instructors to share the students' focus with the course materials. These materials are implicitly present in the form of the professor's own knowledge and explicitly present in the professor's use of physical media like texts and images. Prof. Y, for example, teaches a course in Creative Writing. His goal is to have students make rapid observations and quickly compose descriptions of them. Before discovering that PowerPoint could automate this learning activity, Prof. Y brought video clips of television material (like news, documentary, sitcoms or even commercials) that he had taped at home the night before. He would show a brief clip–perhaps thirty seconds–and then ask students to write in a period of about four minutes everything they had observed. He then explained how the impressionistic written material could be structured into a particular genre, such as a short story. Students would spend the next fifteen minutes shaping their writing into the form Prof. Y just explained. This is an activity that Prof. Y, like Prof. W, already had in his curriculum. Using the technology to help advance it made it better, though, because the video clips could be digitized and played in PowerPoint. Prof. Y experimented with this and discovered an increased versatility: The exact selected thirty seconds could be cued and replayed, like turning the pages of a book, without the inexact forward and reverse searching on the VCR. He could also list the instructions for students to record their observations and have them displayed on the screen during the four-minute impressionistic writing session. He found that he could even have a timer display on the screen to show how much time the students had remaining. Most importantly, he could have half a dozen video options from which students might choose while they engage in a choice-making activity on the method and content of their response. Both video and audio used in interactive ways provide a different kind of learning environment because they recreate the activity and offer the learning community more ways in which to respond to the materials than they would have had otherwise.

Another teaching strategy to increase student interaction is, to have, like Fr. H in an earlier example, students themselves present research findings. In a course entitled Child Psychology, for instance, Dr. D requires students to complete a final term project in which students conduct a literature review on a research topic in child psychology. The students present their reports orally in a simulation of a professional psychology conference. In recent years, more students have requested a screen and data projector so they could present their reports to the class using a PowerPoint document. After noticing that students who prepared PowerPoint presentations for their reports were more likely to have better-organized reports than others, Dr. D decided to require all students to use PowerPoint to give their reports. He created a PowerPoint template for the students that included: title screen, introduction, literature review, results and discussion, conclusion and references. He provided guidelines that students could use in preparing each section, including a sample PowerPoint that contained both the instructions for the assignment and examples of a hypothetical research report. When students give class presentations of their own, of course they are going to rely on the same methods of presentation that have been modeled to them the entire semester. Those who have a familiarity with the program can work with those who are not yet familiar through the instructor's creation of student groups. The synergy created by these groups will cause them to give better class presentations over time as they become more familiar with the software. Better class presentations, it is hoped, will lead to greater facilitated dialogue. As happened in Fr. H's class, the greater facilitation of dialogue led to class presentations developing into panel discussions where those presenting their material actually became leaders of a class discussion concerning those materials. This leads to the technology itself becoming less of a focus. Instead, the ends for which the technology is being used, is the focus. Instructors could internalize this gradual shift in focus by first using presentation rubrics to emphasize technical features and later using them to emphasize content and scope.

By teaching the students how to better package their presentations, the course instructor is actually modeling a learning strategy. The students learn how to author their presentations in the same way the instructor does. The development of that kind of literacy helps them learn the academic content since they invest so much time in the process of organizing it for intelligible delivery to a class of their peers. Because the instructor has already gone through a pedagogical shift in his approach to the students, he or she is no longer lecturing but interacting with the class through a series of learning tools he or she has created. This new literacy opens new opportunities for instructors and students to find practical ways to make use of their study materials in the context of a real discussion. Moreover, they learn the new skill of information filtering. The information age has brought so much content that the only way for learners to deal with it intelligibly is for them to develop the additional skill of packaging the content into easily digestible morsels. Students who can package their lecture materials into manageable and useful units of information have learned far more than those who merely receive large chunks of data from their instructors and end the year without any applications for them. This concept works with the instructor, too—it is far better to learn how to package course content into useful learning modules than it is to try to race through all knowledge in a given subject area in the course of a sixteen-week semester.

Modularization of course materials also serves another practical function since it is more effective to disseminate those materials electronically to each student via course management tools like WebCT or

Blackboard, on the school network, or even through email. If the PowerPoints are designed with the idea of embedded links within each slide that provide students opportunities to engage in choice-making over the areas of a given module that most interest them, then it is quite possible to also embed within the PowerPoints assignments contextualized with the given theme of a specific slide. These assignments can appear in the form of web pages that open when the link is clicked to reveal short readings and response fields that the student can complete for submission to his or her instructor. Upon completion, the student can close the web browser and continue with the slides until he or she reaches a slide in a given presentation that links to a completely new PowerPoint. To make these slides meaningful for the student, the professor might embed audio and video interpretations not only on slide open but also near various pictures that would contextualize the event. Since PowerPoints were not meant to be filled with great amounts of text but to have that text truncated into bullet form, the audio interpretation of those bullets is necessary when a student accesses it on his or her own computer for continued review.

A final future direction might be downloading a free plug-in from Microsoft called Producer that allows the user to capture slideshows within a cinematic timeline so that an audio or video file plays while the slideshow advances on its own. These interactive videos allow students to incrementally pause and engage the various links posted for informational or assessment purposes, and, in doing so, better prepare themselves for the following day of class discussion. Of course, the professor does not have to immediately reach this level on online delivery of modulated materials. If we have shown anything here it is that this engagement with multimedia in the classroom is a process of graduated complexity where it might take several semesters of working through a single course of study before fully developing an electronic set of materials sufficient to sustain an entire learning experience.

This article has illustrated how some instructors have created more interactive teaching and learning experiences using PowerPoint in the classroom. PowerPoint is a valuable tool for classroom teaching when it does not replicate the traditional lecture styles that foster rote memorization over interactive engagement in the learning process. The true transformative power of PowerPoint occurs when instructors change their teaching methodology and move away from stereotypical uses of Power-

Point as a sequential lecture device. This often means instructors learn a new, and ultimately more valuable, way of interacting with their students through the technologies they have available to them.

Footnotes

[1] Free trials of either of these programs can be downloaded from http://www.microsoft.com and http://www.corel.com. While this chapter deals specifically with the use of PowerPoint in the classroom, the ideas presented here easily generalize into use through Corel Presentations.

References

Harris, Daniel, Vic Klimoski, Beverly Lane, and Sebastian Mahfood. (Winter 2003). The Catholic internet: Seminary technologies impacting the teaching and learning environment of the Church. *Seminary Journal*, *9*(3), 80-87.

Hoover, Kenneth H. (1980) *College teaching today: A handbook for postsecondary instruction*. Boston: Allyn and Bacon.

Newble, D. and Cannon, R. (1995) *A Handbook for teachers in universities & colleges: A guide to improving teaching methods*. Third Edition. London: Kogan Page

Orlich, D.C., Harder, R. J., Callahan, R.C., Trevisan, M.S. and Brown, A. H. (2004) *Teaching strategies: A guide to effective instruction*, (Seventh Edition) Boston.

McLuhan, M. (1964) *Understanding media: The extensions of man*. New York: McGraw-Hill.

Nova: Secrets of Photo 51. Producers WGBH/NOVA. Original PBS Broadcast Date: April 22, 2003

Olliges, Ralph and Sebastian Mahfood. (2003). Teaching and learning in the new millennium: Transformative technologies in a transformable world. *Communication Research Trends*, *22*(2), 1-28.

Ong, Walter J. (1982) *Orality and literacy*. Routledge: New York.

Rickman, J. and Grudzinski, M. (April 2000). Student expectations of information technology use in the classroom. *Educause Quarterly*, *23*(1), 24-30.

Quible, Z. K. (June 2002). Maximizing the effectiveness of electronic presentations. *Business Communication Quarterly*, *65*(2), 82-85.

Acknowledgement

The authors would like to thank Dr. Anne Marie Kitz for her pioneering work in the use of PowerPoint as a transactional tool in both face-to-face and online teaching and learning environments.

Integrating Assistive Technology Into Higher Education Settings

Eric Umstead
Murray State University

Tom Lough
Murray State University

Abstract

A description of an innovative way the topic of assistive technology was presented to graduate students in a higher education setting is provided in this chapter. The chapter also shows how the concept of universal design for learning is demonstrated at one university in Kentucky. By utilizing the right software and technology, experts in the field of educational technology are easily crossing distance barriers to provide post-secondary students with information vital for their field of study. Strategies for using the same technology to collaborate with colleagues across campus as well as across the world are also presented.

Introduction

Why include assistive technology in a thorough integration of technology in higher education? According to Kavale and Forness (1996), the number of full-time students entering college for the first time and who reported having a learning disability more than doubled between 1985 and 1996. As a result, higher education classrooms are becoming more and more diverse. Assistive technologies may provide the support necessary for students with disabilities to succeed in postsecondary environments, and possibly improve institutional retention rates for students who might otherwise drop out due to lack of prerequisite academic and/or study skills. The Individuals with Disabilities Education Act (IDEA) of 1997 requires that postsecondary training information be provided during the transition process to students graduating from high schools. Students with disabilities may therefore enter postsecondary environments better informed of their rights to reasonable accommodations in higher education and in the workplace. By integrating various technological resources into the postsecondary curricula, educators will be better prepared to compensate for the variety of individual learning differences they will surely face. One other reason for awareness and integration of assistive technologies connects eventually to our own future. If we are fortunate, we may belong to the population of "temporarily abled" individuals. But eventually, because of accident, disease or just the general wear and tear of old age we will all come to benefit from assistive technologies.

Assistive Technology

According to the IDEA, last reauthorized in 1997, broadly defined, an assistive technology device is:

> ...any item, piece of equipment, or product system, whether acquired commercially off the shelf, modified, or customized, that is used to increase, maintain, or improve the functional capabilities of a child with a disability (Individuals with Disabilities Education Act Amendments of 1997, P.L. 105–17, 20 U.S.C. § 1400 602 (1)).

In other words, assistive technology is developed to improve the ability, performance, and quality of life of individuals with disabilities. Literally thousands of assistive technology devices exist to aid persons with disabilities. These devices range drastically in cost and complexity, from "low tech" pencil grips available at your local discount store to "high tech" alternative and augmentative communication devices, like a nine hundred dollar pocket-sized communication device, and the new tablet personal computers. These extremely portable and versatile personal computers may offer cost effective solutions and benefits to students with a variety of assistive technology needs.

Though having access to the actual assistive technology device is important, access to assistive technology support services may ultimately determine the success of the technological intervention. IDEA offers a federal definition of assistive technology services as:

> ... any service that directly assists a child with a disability in the se lection, acquisition, or use of an assistive technology device.
> Such term includes --
> (A) the evaluation of the needs of such child, including a functional evaluation of the child in the child's customary environment;

(B) purchasing, leasing, or otherwise providing for the acquisition of assistive technology devices by such child;
(C) selecting, designing, fitting, customizing, adapting, applying, maintaining, repairing, or replacing of assistive technology devices;
(D) coordinating and using other therapies, interventions, or services with assistive technology devices, such as those associated with existing education and rehabilitation plans and programs;
(E) training or technical assistance for such child, or, where appropriate, the family of such child; and
(F) training or technical assistance for professionals (including individuals providing education and rehabilitation services), employers, or other individuals who provide services to, employ, or are otherwise substantially involved in the major life functions of such child (Individuals with Disabilities Education Act Amendments of 1997, P.L. 105–17, 20 U.S.C. § 1400 602 (2)).

The preceding definitions provide a foundation for understanding assistive technology devices and the support services necessary to successfully select, acquire, and use such devices to improve the quality of life for individuals with disabilities.

In his booklet *What Every Teacher Should Know About Assistive Technology*, Dave Edyburn (2003) suggests that a challenge exists in educating those who can ensure access to appropriate assistive technology devices and services for all individuals with disabilities. He (Dave Edyburn, 2003) believes "it is essential that every teacher know about the potential of assistive technology and how to access resources that will ensure that children receive appropriate devices and services for overcoming performance problems" (p. 9). The great news for those embracing technology is that assistive technologies have the potential to benefit all students, not just those with disabilities.

Universal Design for Learning

Have you ever used a kerbside ramp designed to make the transition onto a sidewalk easier, or walked through an auto sensing door to enter a building or business? Maybe you've noticed secretaries using speech-to-text software to compensate for carpal tunnel syndrome, or medical professionals using the same technology for dictating medical information. Have you ever participated in an on-line chat session with a friend or colleague? These are just a few examples of architectural and technological innovations that were designed to aid persons with disabilities, but were later found to provide considerable benefit to the general public.

Trent Batson, a professor at Gallaudet University in Washington, D.C., originally designed chat software to enable his hearing-impaired students to have real-time discussions in their writing classes. He believed that if students were actively chatting, they were also collaboratively generating ideas, in addition to reading and expanding upon other students' ideas. The preceding examples illustrate how technologies that were originally designed to assist persons with disabilities became universally beneficial for much larger audiences.

According to the Center for Applied Special Technology (CAST, 2003), Universal Design for Learning (UDL) is a new educational approach that applies recent advances in brain research, in addition to advancing media technologies to aid students with varying individual learning needs in the classroom. The revolutionary aspect of this new paradigm is that all students benefit from the technology, not just one or two special populations. In a UDL environment, it is understood that no single curriculum or software program is able to provide the flexibility needed to effectively address the diverse educational needs of all learners. So an educational setting must be developed that contains a variety of educational and technological resources, software, and Web resources that can be accessed and adapted to meet the diverse learning needs of a variety of students under varying teaching conditions (CAST, 2003). The United States Congress has also added its endorsement of UDL within the last reauthorization of the Assistive Technology Act of 1998. It states that:

> ...the use of universal design principles reduces the need for many specific kinds of assistive technology devices and assistive technology services by building in accommodations for individuals with disabilities before rather than after production. The use of universal design principles also increases the likelihood that products (including services) will be compatible with existing assistive technologies. These principles are increasingly important to enhance access to information technology, telecommunications, transportation, physical structures, and consumer products (Assistive Technology Act of 1998, Section 2 (a) (10) 29 U.S.C. 3001).

Legislation

During the 1970s, 1980s and 1990s, legislation was enacted to aid individuals with disabilities by making assistive technology more accessible. Laws were passed that provide the resources necessary to improve the quality of lives of individuals with disabilities by providing them with expanded access to assistive technology. For eligible students three to twenty-one years of age currently enrolled in public school, the IDEA of 1997 requires the Individual Education Program (IEP) team to consider whether the child with a disability requires assistive technology devices and services to benefit from the individualized education program being developed. Even though the laws and regulations of the IDEA do not follow eligible students into postsecondary environments, these institutions must still make reasonable accommodations for eligible students with disabilities.

There are two significant acts of federal legislation that protect rights, prohibit discrimination, and support the provision of reasonable accommodations for individuals with disabilities. These are the Americans with Disabilities Act (ADA) of 1990 and Section 504 of the Rehabilitation Act of 1973. To qualify as having a disability under either of these statutes, an individual must have "a physical or mental impairment that substantially limits one or more major life activities of such individual, have a record of such impairment, or are regarded as having such an impairment" (Americans with Disabilities Act of 1990, 42 U.S.C. § 12102). Section 504 of the Rehabilitation Act of 1973 mandates that "no otherwise qualified individual with a disability..." will be "excluded from participation in...any program or activity [college, university or other post-secondary institution, or a public system of higher education] receiving federal financial assistance" (29 U.S.C. § 794 et seq.).

As a general rule, institutions must make reasonable accommodations for otherwise qualified individuals with disabilities, unless the accommodations required would present an undue burden on the institution. The ADA places a significant emphasis on the application of reasonable accommodations that enable persons with disabilities to perform essential job functions and enjoy the same rights and privileges as persons without disabilities. The following accommodations may be considered reasonable for students with disabilities at post-secondary institutions: the making of existing facilities used by students readily accessible and usable; the creation of part-time or modified work schedules; acquisition or modification of equipment or devices; the appropriate adjustment or modification of examinations, training materials or policies; and the provision of qualified readers or interpreters, and other similar accommodations (Americans with Disabilities Act of 1990, Section 101 (9), 42 U.S.C. 12101).

Through the careful and appropriate provision of assistive technology, many of the previously stated accommodations could be made effectively, efficiently, and reasonably inexpensively. For example, with continued expansion of on-line course offerings and web-enhanced coursework, many classroom accessibility issues have already been eliminated. In many institutions, faculty members provide study notes, as well as computer-generated presentations (i.e., PowerPoint) on-line or as class handouts for students interested in complementing their own lecture notes. Many textbook publishers are also joining the digital revolution by providing a wealth of supplemental resource materials for both students and faculty on textbook specific websites, making the provision of alternative materials and study aids easily accessible. As speech-to-text and text reader software packages improve and become easier to use, the need for traditional and qualified readers, interpreters, and note-takers may also decrease.

In 1988 the United States Congress stated, within the findings of the Technology-Related Assistance for Individuals with Disabilities Act (1988), that assistive technology devices and assistive technology services were a necessity of life for some individuals with disabilities. The act stated that:

> for some individuals with disabilities, assistive technology devices and assistive technology services are necessary to enable the individuals to have greater control over their lives; to participate in, and contribute more fully to, activities in their home, school, and work environments, and in their communities; to interact to a greater extent with individuals who do not have disabilities; and to otherwise benefit from opportunities that are taken for granted by individuals who do not have disabilities.

After being amended in 1994 and reauthorized in 1998 as the Assistive Technology Act of 1998, this legislation continues its purpose in providing financial assistance to states to strengthen and maintain comprehensive statewide programs of technology-related assistance, for individuals with disabilities of all ages (Assistive Technology Act of 1998, Section 2 (b) (1) 29 U.S.C. 3001).

Innovative Integration Activities

Professors in colleges of education are continually seeking new ways to enhance the education of future teachers. Some educators consider innovations in assistive technology an integral component of comprehensive teacher education. The following example shows in detail how one college professor used a variety of experts in the field of assistive technology to create a number of career illuminating experiences for his graduate students.

For many years, Tom Lough has taught a general education technology course to Murray State University College of Education graduate students. The class includes, among other topics, a section on assistive technology. The assistive technology portion of the course often features a guest assistive technology specialist or a field trip to a local assistive technology center. This aspect of the course is continually refined each year to fully engage all course participants in understanding updated assistive technology concepts, devices, and applications as well as encouraging active participation and interactive dialog.

After reflecting on previous Assistive Technology lessons, an electronic chat component was added to the assistive technology portion of the course. The first part of a two-part plan for the class followed the successful model from earlier semesters. An authority on assistive technology was invited for a two-hour presentation. The second and new component of the course provided student groups with a one-hour online chat session with assistive technology "remote guests" from the United States and Canada. Remote guests were invited based on their expertise, enthusiasm, and desire to help mainstream teachers learn more about assistive technology. The chat session offered students an opportunity to virtually, but in real time, meet professionals who use assistive technology on a daily basis, revisit the assistive technology concepts introduced by the guest speaker, and then reflect on those concepts in terms of their own teaching.

David Daniel, an assistive technology consultant from the Western Kentucky Special Education Cooperative [http://www.wkec.org], made the two-hour live presentation displaying many different examples of assistive technology, including an array of buttons, switches, alternative keyboards, as well as software. He also discussed the innovative use of Tablet PC's in reducing the overall assistive technology costs to public school districts. By using a Tablet PC and a small number of software applications, he has managed to meet the majority of all the districts' assistive technology needs with one reasonably low cost device. For example, this machine could be used at school for a number of educational purposes and then be used at home in the community as an effective and portable augmentative communication device.

He then made a transition to a more future-oriented perspective, demonstrating for the teachers the concept of Universal Design for Learning. He demonstrated Writing with Symbols software to illustrate how to make writing accessible to those who could not read. He informed the teachers, who were the graduate students, that this technology was available now in many of their districts, or would be available in the near future. All many of the teacher participants had to do was ask for the software.

The teachers were then taken into a computer lab and experienced Read and Write Gold, a software application from Ireland that is being used in over 300 schools in Kentucky for assistive technology purposes. Based on UDL, Read and Write Gold provides a number of tools at many levels to help access and compose written material. Some of its features include scanning functions, speech feedback, phonetic spell checker, word prediction, dictionary lookup, and talking scientific calculator. It can be used with many Windows-based word processors, spreadsheets, encyclopedias, web pages, help files, and educational programs.

The chat session portion using the Blackboard courseware [http://www. blackboard.com] followed the Read and Write Gold demonstration. Because the teachers were somewhat familiar with the Blackboard courseware, they believed the chat portion of the courseware was straightforward and easy to use. In addition to supporting regular chat text, the Blackboard courseware also includes a graphic window in which participants could draw interactively, even in color. Since some of the remote guests were unfamiliar with Blackboard, some private practice chat sessions were held with the remote guests a week before the class session. During these sessions, a few minor technical difficulties were observed and eliminated.

The remote guests were asked to facilitate the one-hour session in roughly three parts. In part one, the course participants in each group were to find out something about their assigned remote guest. Each guest was directed to offer information about herself and her professional activity in small increments, but not to offer everything at once. In this part of the chat, the group members became familiar with their guests and established the communication environment. At an appropriate time, the remote guests made a transition to part two, in which they asked the course participants about the two-hour assistive technology presentation. The purpose of this was to revisit the information and help to generalize the concepts for the course participants.

Finally, the remote guests turned the discussion toward the course participants themselves in part three. During the third part of the chat session, course participants reflected on the concept of assistive technology, the potential opportunities available for all teachers, and how continued access to this information may benefit future course participants. One group offered suggestions on how to encourage and provide assistance to "techno-reluctant" teachers in using assistive technology in their own classrooms. Chat sessions also discussed educating classroom teachers to the long-term academic benefits, timesavings, student motivation, and eventual reduction of work associated with systematic technology implementation and training. Regular classroom teachers who were learning about assistive technology provided evidence that word is spreading about this critical application of educational technology. An encouraging message that resounded across the chat groups was that "*assistive technology could and should be used by all teachers throughout the school building, not just special education teachers for special education students*."

During the chat session, the groups of teachers revealed a new understanding of assistive technology. According to the teachers, Daniel introduced them to exceptional software like Symbols 2000, Read and Write Gold, and the accessibility options available in the Windows operating system. Some members excitedly made the connection hoped for by all the experts that participated, that some of the software, such as Read and Write Gold, would be "*excellent for all students, not just those with special needs*." One course participant discovered that her school had just installed the software on computers at her school and she could not wait to get back to school to experiment with it. In addition, course participants were introduced to various switches and keyboards used to aid student access to computer workstations. By combining the alternative methods of access discussed earlier with specially designed software, the session participants began to brainstorm with one another to address technological solutions to academic problems faced by students with disabilities.

Assistive technology may be the answer for the many children expected to grasp content from a textbook written at a level considerably higher than their actual reading level. As teachers become aware that their students' performance on standardized tests may be enhanced by something as simple as digital access to adopted textbooks, assistive technology may assume a greater role in educating all students.

The remote guest chat session was designed in addition to the assistive technology expert speaker to help the classroom teachers taking the course raise their awareness of assistive technology and to increase their internalization of assistive technology concepts. From the remarks during the chat sessions, it appears that this perhaps is beginning to take place. The chat sessions seemed to motivate discussions that were much more assistive technology specific than in previous years. However, there were other immediate indications of this as well. For example, each group delivered a prepared 20-minute presentation addressing assigned dilemmas that focused on issues of equity and ethics. Each group made spontaneous references to assistive technology concepts and activities during their presentations. This indicates that they were already thinking more widely about assistive technology and its implications for them as regular classroom teachers.

According to David Daniel, the best news is that more than 300 of Kentucky's public schools, kindergarten through 12th grade, already have a site license for textreader software installed and available for use. All that is needed are classroom teachers who are interested, motivated, and willing to gather a few new tools for their educational toolbox. So by expanding the technical knowledgebase of teachers in the area of assistive technology, students with disabilities will have additional opportunities for accessing the general curriculum.

As a natural next step to the activities discussed previously and to better prepare future teachers in Kentucky, the Murray State College of Education, in conjunction with the Kentucky Department of Education, obtained a site license for the software Read and Write Gold. Both students and faculty are being trained to use the software. For ease of access, the Read and Write Gold software is installed in both student and faculty computer labs, as well as on faculty laptop computers. The training consists of an assistive technology basics session that defines assistive technology, the concept of UDL, and provides examples of various assistive technology devices and services. This training session also offers a basic understanding of the laws supporting assistive technology, provide basic guidelines as to when assistive technology is warranted, and finally give realistic examples of how to integrate textreader software and other assistive technology concepts into lesson plans to accommodate students with diverse learning needs. Additional training provides hands-on computer time that puts faculty and students behind the keyboard and introduces them step by step to the benefits of Read and Write Gold text reader software, as well as some of the accessibility features embedded in the Microsoft Windows operating system.

Read and Write Gold offers access to high interest material otherwise unavailable to people with learning disabilities. In the past, accessing high interest, low readability print material has not only been difficult for older non-readers, but often came with a stigma attached. Also, individuals with no documented disability, but who struggle to read and write can benefit from this software, which can read text aloud and helps the user to compose written material. As stated earlier, some of its features include document scanning, screen reading, Internet tools, pronunciation facilities, phonetic spell checker, word prediction, a dictionary and even a talking scientific calculator. It can be used with many Windows-based word processors, spreadsheets, encyclopedias, web pages, help files, and educational programs.

Each student is different and learns in unique ways. Knowledge of assistive technology offers future teachers cutting edge tools to help every student succeed. As accountability issues continue to raise concerns over the progress all children are making, the principles of universal design may begin to cause the shift from the traditional to the more universal paradigm. Until the time when we have an UDL standard, we must continue to prepare both faculty and students in institutions of higher education using strategies that can benefit all students.

More frequent and meaningful access to teachers is also becoming a greater concern to parents of public school children with disabilities, university students with disabilities, and potential university students who live in remote areas with no reasonable university access. In an example that meets the access issue head on university faculty are using technology for collaborative purposes as well. Since many non-ambulatory students have difficulty getting from place to place or in some cases class to class, a web camera attached to a computer can bring the professor to the student in cases where getting the student to the professor prove challenging. To model this innovative point to his students, one professor conducted his educational technology class that usually met in a university classroom in Murray, Kentucky, from another classroom at Toyama University in Toyama, Japan. Graduate students in Murray Kentucky used Microsoft's Net Meeting software and an interactive television connection to conduct experiments and communicate in real-time with their Japanese postsecondary peers. Later that same evening, United States and Japanese university students communicated with one another through an Internet chat session as a culminating activity. This activity demonstrated to the participants that collaboration could effectively take place between students and/or entire classes at the same school, across the state, across the country, or even across the planet. Although in this instance an interactive television (ITV) connection gave the video and sound higher quality so that a larger audience could participate in a more meaningful way, a similar session whose video and audio quality may be lower can be replicated by using Microsoft's Net Meeting, a web camera, and a computer with high speed Internet access and a microphone.

Conclusion

One of the resulting benefits from this Murray State University/Toyama University collaborative class experiment is that a number of professors from both universities are now collaborating via web camera and Microsoft's Net Meeting on additional projects ranging in topics from bullying and school violence initiatives to researching new theories in the behavioral sciences. This is just one more example of technology designed for one purpose, being productively utilized for other purposes.

An additional benefit to the university is that in part because of the various collaborative efforts of the faculty, access to new technology, as well as encouragement and incentives by College of Education administration, the faculty in the Murray State University College of Education are becoming more technology literate and are expressing interest in exploring ways technology may be used in the classroom with students and for collaboration with other colleagues. The concept that assistive technology may also be used to improve the university's retention rates by aiding college students with disabilities as well as those students who have had other learning problems in the past is also beginning to be discussed. With the potential to improve university retention and graduation rates, the concepts of Universal Design for Learning and the use of assistive technology may play a greater role on university campuses in the future.

References

Americans with Disabilities Act of 1990, P.L. 101-336, 42 U.S.C. 12101 § 3 (2),

Americans with Disabilities Act of 1990, P.L. 101-336, 42 U.S.C. 12101 § 101 (9)

Assistive Technology Act of 1998, P.L. 105-394, 29 U.S.C. 3001 § 2 (a) (10)

Assistive Technology Act of 1998, P.L. 105-394, 29 U.S.C. 3001 § 2 (b) (1)

Center for Applied Special Technology. (2002). *Universal design for learning*. Retrieved December 29, 2003, from http://www.cast.org/udl/UniversalDesignforLearning361.cfm

Center for Applied Special Technology (2003). *Universal design for learning in classroom practice*. Retrieved December 29, 2003, from http://www.cast.org/udl/index.cfm?i=191

Day, M., & Batson, T. (1995). The network-based writing classroom: The ENFI idea. In Collins, M & Zane B (eds.), *Computer mediated communication and the online writing classroom - Volume two: Higher education*. (pp. 25-46) Cresskill, NJ: Hampton Press

Edyburn, D. (2003). *What every teacher should know about assistive technology*. Allyn and Bacon, Boston, MA.

Kavale, K. A., & Forness, S. R. (1996). Learning disability grows up: Rehabilitation issues for individuals with learning disabilities. *The Journal of Rehabilitation*, *62*, 34–42.

Individuals with Disabilities Education Act Amendments of 1997, P.L. 105–17, 20 U.S.C. § 1400 602 (1).

Individuals with Disabilities Education Act Amendments of 1997, P.L. 105–17, 20 U.S.C. § 1400 602 (2).

Lough, T., Umstead, E.W. (2002). How to bring assistive technology to life in a university educational technology course? Invite remote guests! *The Catalyst*, *19*(2), 6-8.

Technology-Related Assistance Act Amendments of 1994. P.L. 103-218 § 2 (3)

Umstead, E.W. (2003). Assistive technology: Not just for special educators. *The Catalyst*, *19*(4).

Integrating Multimedia Technology and Interactive Video into Counseling Skill Training

Leping Liu
University of Nevada, Reno

Paul C. Abney
University of Virgin Islands

Abstract

This chapter presents a successful experience of integrating multimedia technology and interactive video into counseling skill training, focusing on the designs and implementation of the integration in a graduate level counselor practicum course. The design of instructions and procedures for counseling skill training is based on the Skill Counselor Training Model, and the design of the integration of interactive video is based on the Technology Integration Design Model. A video editing program iMovie was used to create multiple formats of information, and develop synchronous interactions in a counseling skill demonstration interactive video project.

Introduction

Technology has been integrated into the practice of counseling since its original applications in the 1940s (Berry, Srebalus, Cromer, and Takacs, 2003; Granello, 2000; Myrick, 1997; Neukrug, 1991; and Walz, Harris-Bowlsbey, Dinkmeyer, Carlson, & Nelson, 1990). Furthermore, technology has been infused into instruction and training of higher education students enrolled in Counselor Education programs over the past two decades (Harris-Bowlsebey,1984; Cabaniss, 2002; Casey, 1999; Neukrug, 1991; Sampson, Kolodinsky, & Greeno, 1997; and Watson, 2003). Previous literature has exhibited research and practice at the initial level of technology integration–what to use, exploring appropriate technologies for counseling education (Bassoppo-Mayo, 1997; Sturges, 1998; and Bornas, Tortella-Feliu, Llabres, & Fullana, 2001). The current chapter will present a successful experience at a higher level of integration–teaching with technology (Maddux, Johnson, & Willis, 2001; and Liu & Johnson, 2003), focusing on the design and implementation of integrating interactive video technology into a counseling skill training graduate course. The design of instructions and procedures for counseling skill training is based on the Skill Counselor Training Model (Downing, Smaby, & Maddux, 2001; Smaby, Maddux, Torres-Rivera, & Zimmick, 1999; Zimmick, Smaby & Maddux, 2000), and the design of the integration of interactive-video is based on the Technology Integration Design Model (Liu & Velasquez-Bryant, 2003; Liu & Johnson, 2003a), using the iMovie video editing program.

The Counseling Skill Training Course

Counselor Education is "a process that prepares counselors in both didactic and clinical aspects of counseling" (Council for Accreditation of Counseling and Related Educational Programs 2001 Standards). The development and refinement of counseling skills has long been a focus of counselor training in higher education (Carkhuff, 2000; Daniels, 2003; Egan, 2002; and Ivey, & Ivey, 2003). The counseling skill training course introduced in the current chapter is titled "Counseling Practicum," a 3-credit, 45-hour, semester-long course required by the School and Community Counseling masters degree program in a western state university.

Course Content

This course is based on providing practicum instruction that focuses on the development of students' basic counseling skills. In turn, the development of these skills provides students with the knowledge and practice necessary to begin counseling effectively. Students then learn to systematically apply specific counseling skills according to the stage of counseling that they have reached with their clients. Based on their knowledge and practice, students formulate personalized counseling theories, methods, and styles.

Course content areas are based on the CACREP (Council for Accreditation of Counseling and Related Educational Programs) 2001 Standards, within a scope of the following topics: (1) The evolution of professional counseling, (2) Professional identity of community counselors,

(3) Practice settings, (4) Ethical and legal issues, (5) Developmental foundation, (6) Client assessment, (7) Environmental and needs assessment, (8) Community counseling with individuals, (9) Community counseling with families, (10) Community counseling with groups, (11) Educating and programming, (12) Consultation and supervision, (13) Case management and coordination of resources, (14) Policy formation and advocacy, (15) Evaluation of services, and (15) Future directions. Required counseling skills are infused under this scope.

Counseling Skills

In the Counselor Practicum course, students are expected to develop eighteen specific counseling skills through three different stages of the counseling processes, along with two related objectives in each stage.

Exploring Stage: The objectives of this stage are: (a) Attending, and (b) Questions and Reflections; and the following counseling skills are practiced: (1) eye contact, (2) body language, (3) verbal tracking, (4) questioning, (5) paraphrasing, and (6) summarizing.

Understanding Stage: The objectives of this stage are: (a) Interchangeable Empathy, and (b) Additive Empathy and the following skills are practiced: (7) feeling and content, (8) self-disclosure, (9) concrete and specific, (10) immediacy, (11) situation action and feelings, (12) confronts caringly.

Action Stage: The objectives of this stage are: (a) Decision Making, and (b) Contracting and the following skills are practiced: (13) deciding, (14) choosing, (15) consequences, (16) agreements, (17) deadlines, and (18) reviewing goals and actions to determine outcome.

Proposed by Downing, Smaby, and Maddux (2001), the eighteen skills are highly structured along with the skill training process. Counselor trainees will first learn, assess and master basic attention and reflection skills–such as eye contact and summarizing client's statements. They will then apply intuitive skills such as empathy, self-disclosure and confrontation. Finally they will develop cognitively complex decision-making skills such as deciding and choosing.

Course Requirements

The Counseling Practicum course requires counselor trainees/participants to: (a) perform a minimum of 40 hours of direct service with clients, (b) receive a minimum of one hour per week of individual supervision (using audiotape, videotape, and/or direct observation), (c) receive a minimum of one and one-half hours per week of group supervision with other students in similar practica, and (d) receive evaluation of their performance throughout the practicum including a formal evaluation at the completion of the practicum. The formal evaluation of skills is based upon the instructor's evaluation of students' Skill Training Demonstration Tapes. Students record fifty-minute counseling sessions demonstrating the eighteen skills.

Process of Counseling Skill Training

Counseling skill training in this course progressed through two levels: (1) acquisition of counseling skills and (2) application and retention of counseling skills. At the first level, students learned basic knowledge of the skills and principles to apply them. At the second level they applied and improved their skills through counseling sessions.

Level I: Acquisition of Counseling Skills

Instructional design of counseling skill learning was based on the "Skill Counselor Training Model" (Ivey & Ivey, 2003), which is a practice oriented learning model for single skill training with a step-by-step progression. Students learned each single skill through the five dimensions of the model: (1) orientation, (2) observation, (3) reading, (4) practicing, and (5) self-assessment (Ivey & Ivey, 2003).

Orientation of the Skill: Orientation of skill is focusing on a single skill, a vital part of the holistic interview. Students obtained general knowledge of the skill from a PowerPoint presentation on overall structure of the eighteen skills, specific functions of a single skill, general principles of applying it in a session, and the scale to measure the mastering of the skill.

Observation of the Skill: Observation of a skill in action could be done through a transcript of an interview, an audiotape, videotape, or a live demonstration. Students watched a six-minute videotaped session in which an experienced counselor/instructor demonstrated the skills with a volunteer "client." In addition, the class instructor, or graduate assistant, provided a "live" demonstration of the skills that were assigned for each class period using a student volunteer. By observing the demonstration, students gained a picture of and concrete ideas about how that skill could be used.

Reading About the Skill: Reading about the skill was a process to digest and refine what they learned from *orientation* and *observation*. Students reviewed the main points of effective usage of the skill and formulated initial ideas for their practice.

Practicing the Skill: Practicing the skill was performed in roleplays in class. Students were assigned to, and formed training groups. Each member, in turn, acted as a counselor and attempted to perform the assigned set of skills identified in the model.

Self-assessment: Self-assessment was implemented within the training group. Students were instructed to provide critical feedback to the acting counselor. After receiving feedback, the counselor/trainee, with guidance from other group members, attempted to demonstrate skills that did not occur. The class instructor assisted those who continued to have difficulty demonstrating these skills, and they cued the counselor/trainee until he or she was successful.

As described above, the eighteen counseling skills were developed through three stages of counseling processes–exploring, understanding, and action. The same five-dimensional learning process was repeated for all eighteen skills. After they learned basic knowledge and effective usage of the six skills in one of the three stages, they moved to the next level of training–applying the skills. They were required to provide direct service with clients through counseling sessions and demonstrate the six skills. The process of first learning the skill and then applying it was repeated for all three stages.

Level II: Application and Retention of Counseling Skills

At the application level of counseling skill training, students were required to produce skill training demonstration videotapes through their counseling sessions. Students completed demonstration videotape projects by performing the following steps:

- recording the entire counseling session
- identifying interactions between the client and the counselor trainee (client's statement or action, and counselor trainee's response)
- identifying the skills demonstrated in each interaction (use and rationale for use of the skill)
- providing the counselor trainee's self-critique (thoughts and alternative response)
- importing the instructor's feedback.

It is believed that this process assists the counselor trainees to increase discrimination skills, to identify strengths and weaknesses (including unresolved personal issues that may be systematically hindering their progress) and to point out opportunities for theoretical application and integration (Baumgarten & Roffers, 2003). In addition, these practice-oriented and step-by-step progressive activities can serve as a bridge between the trainees' training experiences and subsequent professional efforts, facilitating skill transfer and retention.

Transcribed Manuscript

In the past, a traditional *Transcribed Manuscript* was used for the videotape project. The counselor trainees (students in the Counseling Practicum course) created *written scripts* for steps 2, 3, and 4, of the demonstration video project and the instructor provided *written feedback* for each skill or a set of skills.

While producing the manuscripts, students first identified and labeled a client's expression as a stated belief, request, concern, interest, and intent. Then they retroactively created the best possible interchangeable response, even if it was different from their original response. Students then wrote and compared their original response with the retroactively created response. The following is an example of the written manuscripts:

Tape Location: This interaction is located at 112 on the tape.
Interaction 1:

> Client: "I have been so angry for so long I just don't know how to manage my angry feelings."
> Counselor: "Angry?"

Use and rationale: This response reflects the use of *verbal tracking*.
Thoughts: At the time that I made this statement I was thinking that I would prefer to use a question or other exploratory response. However, I was feeling pressure to practice the verbal tracking. At this moment I was thinking that verbal tracking did not seem nearly appropriate as other responses might be.
Alternative response: "How does your anger seem unmanageable?" The use of more open-ended questions might have elicited more information. I felt that this would have been more beneficial in helping the client with her anger.
Instructor's feedback: (comments and suggestions from the instructor)

For all the identified interactions on the tape, the same format was used to produce the manuscripts. Students in the course were required to produce four skill demonstration tapes. Each of the first three tapes demonstrated six skills on each of the three stages, and the fourth one demonstrated all eighteen skills.

After completing the manuscripts for one counseling session, students reviewed their tapes with the self-critique and instructor's feedback. From the review, they refined their understanding of the skills and formulated better ideas of using them for further practices.

Weaknesses of Using Transcribed Manuscripts

Compared with research findings regarding the effectiveness of learning, using transcribed manuscripts in the skill demonstration tape does have two weaknesses. First, the evaluation comments to students' skills (students' self-critique in the written manuscripts, and the instructor's written feedback) are text-only information, while previous research suggests that students would learn more effectively when they are exposed with information from multiple sources with various formats such as text, sound, graphics, animation and still or moving pictures (Counts, 2004; Liu & Johnson, 2003b; Heinich, Molenda, Russell & Smaldino, 2002; and Alessi & Trollip, 2001; and Liu, 1997). Second, the communication between the instructor (feedback comments) and students (review of skill performances demonstrated in the tape) is asynchronous, while the literature shows that students are more motivated to learn and therefore would learn better when synchronous communications take place between the instructor and students (Liu, 2003; Liu, & Maddux, 2003; Vonderwell, 2003; Hewitt, 2003; Ohlund, Yu, Jannasch-Pennell & Digangi, 2000; and Swan, Shea, Fredericksen, Pickett, Pelz & Maher, 2000).

Accordingly, it is believed that students' counseling skill training could be improved if (a) feedback is provided in multiple formats, and (b) synchronous interactions between the skill performance and feedback (both self-critique and instructor's comments) are merged into the skill demonstration tape. To fulfill this, a new method of using multimedia technology to produce interactive video has been adopted into the Counseling Practicum course.

Integration of Interactive Video

In this course, the integration of multimedia technology and interactive video was implemented within a three-dimension technology integration model (Liu & Velasquez-Bryant, 2003; and Liu & Johnson, 2003a). A video-editing program *iMovie* was used at the second level of counseling skill training–*Application of the Skills*, to produce an interactive skill demonstration videotape.

The Three-Dimension ITD Integration Model

Based on a review of research and practices over the past three decades, Liu and Velasquez-Bryant (2003) proposed a three-dimensional (3D) technology integration model. The first dimension, *information (I)*, represents the learning or teaching of the content, and any supporting resources and materials. The second dimension, *technology (T)*, represents the hardware and software tools that can be used appropriately to support or enhance learning and teaching. The third dimension, *instructional design (D)*, represents a set of rules for instructional design. This 3D-ITD Model illustrates a merging of the uses of information and technology into a system that follows the rules, theories, and models of instructional design. It is assumed that technology-based learning only occurs as the result of the *integration* of all three dimensions, namely Information (I), Technology (T), and Instructional Design (D). No single dimension or any combination of just two dimensions in this model will produce effective learning.

In the counseling training course described above, the merging of *information* and *instructional design* was completed at both levels of counseling skill training. The eighteen counseling skills could be viewed as *information* (learning contents). The instructional design used at the first level (*Acquisition of Counseling Skills*) was the five-dimension *Skill Counselor Training Model*, and at the second level (*Application and Retention of Counseling Skills*) was the five-step procedure to produce skill demonstration tapes. This well-structured course procedure illustrated the merging of two dimensions in the ITD model — Information (I), and Instructional Design (D). To complete the 3D integration model, the third dimension Use of Technology (T), the video editing program iMovie was integrated into the course.

Process of Using iMovie

iMovie is a video editing software program. Video editing is the process of assembling video clips (segments of the video), still images, and audio into a finished package (Heid, 2003). iMovie has the major functions that enable us to create interactive video. First, using iMovie, we can import digital video from a digital video camcorder using a FireWire connection. iMovie controls the camcorder during the importing process, stashing incoming clips on its Clips pane (also called Shelf Palette). Second, after the video is stored as a set of clips, we can edit clips, optionally recording voice, adding music, labeling titles, or creating effects and scene transitions. Third, the most useful function of iMovie is that we can select and add clips into the movie, and rearrange the sequence of clips.

The use of iMovie was designed to meet the objectives of counseling skill training, to follow the procedures of skill learning and practicing, and to compensate for the weaknesses revealed in traditional methods of transcribed manuscripts.

As described in a previous section, while producing skill demonstration video, students followed the same steps: (1) recording the counseling session, (2) identifying interactions, (3) identifying skill(s), (4) importing self-critique, and (5) import instructor's feedback. Instead of using written scripts, iMovie was used in steps two through five to create multiple formats of feedback, and synchronous interactions in the skill demonstration tape.

Creating Digital Video: In the first step, the sessions were recorded using either digital (DV) or analog (VHS) videotapes. If analog, the tapes were converted to digital format, getting them ready to be imported into iMovie.

Identifying and Assembling Video Clips: In steps two and three interactions and skills were identified, by selecting appropriate video clips. In order to identify single skills from the video, students first reviewed the original videotape of the session and noted the time code location where the individual skill was demonstrated. They then imported the digital video into iMovie, loading only the skill segments. The skill segments were saved as video clips and imported into the Shelf Palette, which is the default location in iMovie to import video clips. When saving the clips, editing options were available to add titles, labels or descriptions of the skill to the clips.

After the skill video-clips were saved and labeled, students then dragged them to the Timeline Viewer in the sequence they wanted the skills to be demonstrated. Timeline Viewer is the place where an iMovie project is assembled. The Timeline Viewer places clips in a graphical view and students can rearrange the order of the clips by simply dragging and dropping the clips.

Using the Timeline Viewer, interactions between the counselor (the student) and the client, and counseling skills the counselor demonstrated through the counseling session were identified from the original videotape and reassembled to show only the individual skills. Text-based information was used to label and describe the skills.

Adding Sounds to the Clips and Exporting a Movie: In the fourth step, students added self-critique audio tracks at the end of each skill, indicating how they felt the skill was or was not adequately demonstrated. In the Timeline Viewer, there are two separate sound tracks that can be manipulated. The track that shows "voice 01" and "voice 02" is the place students recorded and added their voices. Students were able to add the voice of their own self-critique as audio tracks to any clip. They then continued adding self-critique audio tracks to other clips in the Timeline Viewer until they completed all the skill clips. Using the sound track, students' self-critique was created in an audio-format.

At this point, all clips need to be exported or published as one movie so that it can be viewed outside the iMovie system. The completed video projects were then exported (downloaded) from the computer back to digital video and to an analog VHS, and both tapes were submitted for review to the supervising instructor.

Creating Instructor's Feedback Video: This part was the instructor's work. The instructor, while observing the VHS version of the iMovie projects, provided videotaped feedback on the digital tapes commenting on the strengths and weaknesses of each skill demonstrated by the participants. Now, the digital tape included skill-based interactions with the student's self-critique, and the instructor's feedback at the end. In this way, instructor's feedback was created in a video-format.

Creating Interactive Video: In the fifth step, the digital tapes containing instructor feedback were returned to the students. Using iMovie, the students uploaded the instructor's feedback into the computer video project, and then repeated the procedures in previous steps, adding the instructor's feedback-clips to the end of each skill-demonstration clip, which provided a comprehensive assessment, both self and supervisory, of the counseling skills demonstrated by the student. The finished video presentation consisted of: (a) a title and skill introduction, (b) a demonstration of the micro-skill technique employed, (c) audio self-critique by the counselor (the student), and (d) audio and video feedback by the instructor.

This interactive video created with iMovie compensated for the two weaknesses of using transcribed manuscripts. First, it provided feedback in multiple formats (text, sound, and video). Second, when students reviewed this video, they could receive feedback from the instructor "immediately" after the skill demonstration, almost as if they were communicating synchronously with the instructor.

Students' Responses

Since our experiences were initial attempts to explore the use of interactive video in counseling skill training, we carefully observed students' performances during the process of the video project, and collected their responses to the use of the new technology. After two semesters of employing the video editing technique, students' responses could be sorted into two categories: (1) attitudes toward the two methods–written manuscripts and using a video editing software, and (2) necessary support for students to use the new technology.

Almost all students felt positive about using the video editing technique. They realized that, compared with the traditional method of transcribed manuscripts, the multiple formats of feedback and synchronous interactions created in the demonstration tape did help them more in understanding and improving their skills. The addition of original artwork, music and artistic perspective to each presentation gave a personal touch and provided a creative element unique to each student. The results from the course survey also showed that students enjoyed learning, and were motivated to create the interactive video. Students also showed various levels of anxiety depending on their technology skill level and previous experiences with technology. In our Counselor Education program, most students only mastered basic computing skills, and rarely touched multimedia applications.

According to feedback from the students:

(1) a well devised training program that includes everything from camera use to software instruction to final tape production is crucial.
(2) technology assistance in the form of personnel available to help the students in the editing facility is invaluable.
(3) access to the video lab must be made readily available as the various work schedules of students prohibit editing during limited scheduled times.

Discussions and Conclusions

We want to conclude this chapter with some thoughts from our experiences. First, technology integration starts with the need to improve teaching and learning, and it focuses on the instructional design of the learning content. Technology is not a standalone tool and its use should be integrated into the learning process, as illustrated in the 3D-ITD model. In our experiences of integrating interactive video into the counseling skill training, all the procedures of creating the interactive video are a part of the skill learning and training and the product created using the video editing software should be the skills students mastered rather than the interactive video itself.

Second, technology-based learning is a learner-centered learning process. As in this Counselor Practicum course, students learned from analyzing and summarizing the skills in the process of creating the skill demonstration interactive video, which is consistent with a constructivist approach of "learning from doing."

Third, technology-based learning would be more effective if it focuses on the progressive process of learning (Liu & Maddux, 2003). The use of iMovie did develop such a learning experience. After each step of the video project–identifying the skill clips, self-critique, editing and re-editing the movie, students gained more understanding about the skills. Students learned from the whole process rather than just "one skill at one step."

Fourth, technology-based learning would be more creative if it is designed with an open-ended process. For example, the skill video could be viewed as an open-ended project, since students can always improve their work through.

We have presented the design and implementation of the integration of interactive video technologies into counseling skill training. We view this effort as an initial step and recognize the limitations of our designs and experiences. We hope that other educators and researchers can use this first step in designing further studies and technology integration practices.

References

Alessi, S. M., & Trollip, S. R. (2001). *Multimedia for learning: Methods and development*. Boston, MA: Allyn and Bacon.

Bassoppo-Mayo, T. C. (1997). Digital video interaction technology: The world's new frontier in education, training and development? *International Journal of Instructional Media, 24*(4), 305-314.

Baumgarten, E., & Roffers, T., (2003). Implementing and expanding on Carhuff's training technology. *Journal of Counseling & Development, 81*(3), 285-291.

Berry, T., Srebalus, D. J., Cromer, P. W., & Takacs, J. (2003). Counselor trainee technology use, skills, learning styles, and preferred modes of instruction. *Journal of Technology in Counseling, 3*(1). Retrieved February 12, 2004, from http://jtc.colstate.edu/vol3_1/Takacs/Takacs.htm

Bornas, X., Tortella-Feliu, M., Llabres, J., & Fullana, D. (2001). Computer assisted exposure treatment for flight phobia. *Psychotherapy Research, 11*(3), 259-273.

Canabiss, K. (2002). Computer-related technology use by counselors in the new millennium: A Delphi study. *Journal of Technology in Counseling, 2*(2). Available: http://jtc.colstate.edu/vol2_2/cabaniss/cabaniss.htm

Carkhuff, R. (2000). *The art of helping* (8th ed.). Amherst, MA: Human resources Development Press.

Casey, J. A. (1999). Computer assisted simulation for counseling training of basic skills. *Journal of Technology in Counseling, 1*(1). Available: http://jtc.colstate.edu/vol1_1/simulation.htm

Council for Accreditation of Counseling and Related Educational Programs. (n.d.). 2001 Standards. *Counselor Education* (Glossary section). Retrieved February, 12, 2004, from http://www.cacrep.org/2001Standards.html

Counts, E. L. (2004). *Multimedia design and production for students and teachers*. Boston, MA: Pearson Allyn and Bacon.

Daniels, T. (2003). A review of research on microcounseling: 1967-present. In A. Ivey & M. Ivey (2003). *Intentional interviewing and counseling: An interactive CD-ROM.* Pacific Grove, CA: Brooks/Cole.

Downing, T. K., Smaby, M. H., Maddux, C. D. (2001). A study of the transfer of group counseling skills from training to practice. *Journal for Specialists in Group Work, 26*, 156-167.

Egan, G. (2002). *The skilled helper* (7th ed.). Pacific Grove, CA: Brooks/Cole.

Granello, P.F., (2000). Historical context: the relationship of computer technologies and counseling. *Counseling and Personnel Services* (CGO 30192), 1-11.

Harris-Bowlsebey, J., (1984). High touch and high technology: The marriage that must succeed. *Counselor Education and Supervision, 24*(1), 6- 16.

Heid, J. (2003). *iMovie* at a glance. *MacWorld, 20*(4), 10-12.

Heinich, R., Molenda, M., Russell, J. D., & Smaldino, S. E. (2002). *Instructional media and technologies for learning*. (7th ed.). Columbus, OH: Merrill Prentice Hall.

Hewitt, J. (2003). How habitual online practices affect the development of asynchronous discussion threads. *Journal of Educational Computing Research, 28*(1), 31-45.

Ivey, A. E., & Ivey, M. B., (2003). *Intentional interviewing and counseling: Facilitating client development in a multicultural society.* Pacific Grove, CA: Brooks/Cole.

Liu, M. (1997). The effects of HyperCard programming on teacher education students' problem-solving ability and computer anxiety. *Journal of Research on Computing in Education, 29*(3), 248-262.

Liu, L. (2003). *Communication design for online courses: Effects of a multi-layer approach.* Presented at E-Learning Word Conference, Phoenix, AZ, November 7 -11, 2003.

Liu, L., & Maddux, C. (2003). *Online course design and research.* Presented at E-Learning World Conference, Phoenix, AZ, November 7 - 11, 2003.

Liu, L., & Johnson, L. (2003a). *A new approach of design: Technology integration as information system development.* Presented at: SITE (*Society for Information Technology and Teacher Education*), Albuquerque, NM, March 28, 2003.

Liu, L., & Johnson, L. (2003b). *Structure-based versus function-based method in multimedia courseware design.* Presented at: SITE (*Society for Information Technology and Teacher Education*), Albuquerque, NM, March 28, 2003.

Liu, L., & Velasquez-Bryant, J. (2003). An information technology integration system and its life cycle: What is missing? *Computers in the Schools, 20*(1-2), 93-106.

Maddux, C. D., Johnson, D. L., & Willis, J. W. (2001). *Educational computing: Learning with tomorrow's technologies.* (3rd ed.). Boston, MA: Allyn and Bacon.

Myrick, R. D. (1997). Traveling together on the road ahead. *Professional School Counseling, 1*(1), 4-8.

Neukrug, E. S. (1991). Computer-assisted live supervision in counseling skills training. *Counselor education and Supervision, 31*(2), 132-139.

Ohlund, B., Yu, C. H., Jannasch-Pennell, A., & Digangi, S. A. (2000). Impact of asynchronous and synchronous internet-base communication on collaboration and performance among K-12 teachers. *Journal of Educational Research, 23*(4), 405-420.

Sampson, J. P., Kolodinsky, R. W., & Greeno, B. P. (1997). Counseling on the information highway: Future possibilities and potential problems. *Journal of Counseling and Development*, *75*(3), 203-212.

Smaby, M. H., Maddux, C. D., Torres-Rivera, E., & Zimmick, R. (1999). A study of the effects of a skills-based versus a conventional group counseling training program. *Journal for Specialists in Group Work*, *24*, 152-163.

Sturges, J. W. (1998). Practical use of technology in professional practice. *Professional Psychology*, *29* (2), 183-188.

Swan, K., Shea, P., Fredericksen, E., Pickett, A., Pelz, W., & Maher, G. (2000). Building knowledge building communites: Consistency contact and communication in the virtual classroom. *Journal of Educational Computing Research*, *23*(4), 359-383.

Vonderwell, S. (2003). An examination of asynchronous communication experiences and perspectives of students in an online course: A case study. *The Internet and Higher Education*, *6*(1), 77-90.

Waltz, G. R., Harris-Bowlsebbey, J., Dinkmeyer, Jr., D, Carlson, J., Nelson, R. C., Krockover, G. H. (1990). A technological world. *Counseling and Personnel Services* (CGO22269), 1-13.

Watson, J. C. (2003). Computer-based supervision: Implementing computer technology into the delivery of counseling supervision. *Journal of Technology in Counseling*, *3*(1). Retrieved February 12, 2004, from http://jtc.colstate.edu/vol3_1/Watson/Watson.htm

Zimmick, R., Smaby, M. H., & Maddux, C. (2000). Improving the use of a group counseling scale and related model to teach theory and skills integration. *Counselor Education and Supervision*, *39*, 248-295.

Teaching Disaster Management: An IT-Enhanced Hybrid Problem-Based Learning and Lecture Model

Adrian Bromage
Coventry University

El Parker
Coventry University

Abstract

An approach to teaching Disaster Management is described that is based upon a hybrid of problem-based learning (PBL) and lectures, supported by the use of a virtual learning environment (VLE). It is implemented in three course units that span a three-year undergraduate degree programme in the UK. As the students' initially lack experience of PBL and using the VLE, the PBL component increases progressively across the three years and the VLE is used in increasingly sophisticated ways in support. Measures to inhibit final year students from plagiarising assessed work from each other are described. Data generated automatically within the VLE indicate that while the students are making use of online learning materials, they presently contribute little to the online discussion forum. It is thought that carefully structuring the latter's use within the module's assessment frameworks will help to change this. The approach offers the benefits of inducting students into professional practices 'hands on' through 'authentic' tasks, while affording the module tutor an explicit context, the lecture component, in which to share their own expertise and experience. It also yields an ever-increasing resources base for students, an advantage when there arguably exist no suitable core texts for the second and third year. However, it raises issues related to University administration systems, and the students' Communication & Information Technologies (C&IT) skills, harnessing students' accumulated knowledge, their motivation and time management. In terms of future developments, the role of the VLE in making the modules accessible to students with disabilities is discussed.

Introduction

This chapter documents the progressive curriculum development of three course units in Natural Hazards that form a theme within several undergraduate programmes run in the UK by Coventry University's Disaster Management subject group. While of particular interest to curriculum developers working within the discipline, is envisaged that the chapter will be of interest to others who would like to develop Problem Based Learning (PBL) in combination with the use of a Virtual Learning Environment (VLE) within the context of undergraduate education.

The course units in question are mandatory core modules, one for each of the three years (Levels) of study for degrees in International Disaster and Engineering Management (IDEM), Development and Health in Disaster Management (DHDM), Emergency and Disaster Management (EDM) and Environmental Hazards in Disaster Management (ENVDM). Students outside the Disaster Management group may also complete them as 'free choice' modules. Over the last three years, class sizes for Levels One, Two and Three have been on average around 64, 34 and 22 students respectively. In a typical year, up to 10 of the students in each cohort might be from other subject areas such as Geography, Tourism Management and Ecology.

The characteristics of typical Disaster Management students have exerted a strong influence on the development of the teaching and learning strategies deployed in the modules. Few Undergraduate programmes of this kind presently exist in the UK, so the students tend to have actively chosen the courses. As a result they arrive with fairly high academic qualifications compared to students of other Level One programmes at the University. The eighteen year olds progressing from colleges of further education often have a long-term interest in charity work, outdoor activities or sports. There are also significant numbers of mature students who tend to be returning to study after taking a few years out to travel and work abroad, or who seek a career change after serving in the armed services, emergency services or medical professions. Overall, the students tend to have high degree of both intrinsic and extrinsic motivation, and this is reflected in high rates of participation in lectures and strong demands of their tutors. The vast majority of the students aspire to gain employment in the fields of disaster, development or emergency management, and over two thirds fulfil this aspiration.

The Communication & Information Technologies (C&IT) That Have Been Integrated

In 1999 Coventry University adopted WebCT institution-wide as its Virtual Learning Environment (VLE). While originally developed for the provision of online learning, the intention was that it be used at Coventry primarily to support traditional, campus-based programmes. The University automatically creates a WebCT site for every module. Each site comprises a standard template that includes hyperlinks to the University Library and its Catalogue, the University Study Skills Centre, the relevant academic departmental homepage and module information including past exam papers. The use made of WebCT beyond this minimum, in terms of adding course content and devising learning activities, is left to tutors' discretion (Deepwell and Syson, 1999). Arguably, the facilities built into the VLE, in particular its electronic 'discussion forum' and internal email, 'module mail', have the potential to increase both the 'contact' and guidance that a tutor can provide for students. They also facilitate the implementation of alternatives to traditional 'chalk and talk' teaching situations, and small group workshop activities supported by consistent software. The VLE is thus well suited to support a PBL approach.

By 2003, almost every personal computer on the campus, for both staff and students, had been connected to the Internet and a dial-up connection service offered for those working off-campus. This has opened up interesting possibilities. The majority of students seem to be fascinated by large-scale disasters caused by natural phenomena which wreak high levels of death and destruction. There exists a wealth of freely available high-quality case study material that is available on the Internet, for example videos, photographs, newspaper articles, which can thus be used to draw their interest. It will be seen in a later section of this chapter that this is particularly important for the Level Two and Three modules.

Overview of the Curriculum in the Modules Under Discussion

The aim was to incorporate into the modules simulations of the kind of problems that might be faced by a disaster manager in real life, assessing hazards and reducing risks. Such problems cannot be solved during a single tutorial session by using an existing 'formula.' Appropriate learning activities must reproduce the real-life activities of sourcing, requesting and receiving data, assessing its quality, analysing and presenting results, then devising a strategy to reduce risk and assessing its appropriateness. It will be argued that problem based learning (PBL) is a model of teaching and learning which matches these requirements, although it will be seen that circumstances militate against a 'pure' PBL approach.

Problem based learning is defined by Savin-Baden (2000) as an approach whereby curriculum content is organised around problem scenarios. PBL seeks to reflect 'real life': real data, realistic scenarios, real time frames and constraints. Together the students explore assumptions and misconceptions and the workability of various strategies, and identify the information and skills needed to devise a solution. This process requires them to identify their own learning needs and pre-existing beliefs, to research, evaluate and critique information, and to test various approaches and identify inappropriate avenues. It is a more flexible and diverse approach than lecture-based formats: there are no 'right' answers, but many options. The lecturer is not viewed as 'all knowing,' and discussions, seminars and tutorial sessions provide a format in which to iron out any problems, misconceptions or errors in the spirit of joint learning. Savin-Baden (2000) contrasts this with 'problem-solving' leaning. In this approach the problems are bounded both by the subject or disciplinary area and by curriculum content. A tutor sets questions on the content that they covered during a lecture, and the students discuss their answers in a subsequent seminar. The focus is on attaining the answers that are expected by the tutor and rooted in the content supplied.

Problem-based learning was originally popularised by Barrows in the context of medical education in the 1960's (Barrows and Tamblyn, 1980). Newman (2003) argues that the theoretical underpinnings of PBL in this context were derived from cognitive psychology, while in its recent adoption in other disciplinary contexts they have been derived from other theoretical positions, such as reflective practice or Vygotsky's (1930) social constructionist position. Savin-Baden (2000) holds that the increased uptake of PBL in higher education over the past 30 years has been driven by calls for a greater emphasis developing competencies and less on 'liberal education', which values knowledge in itself, arguing that PBL is a format that can combine both facets. However, Savin-Baden (2000) is critical of recent PBL literature, perceiving it to have a narrow focus on factors related to its implementation, while neglecting both its impact on learners and institutions and how institutional contexts influence PBL. Similarly Newman (2003) argues that there exists little high-quality and robust evidence to substantiate the claimed advantages of PBL compared to 'traditional' teaching. They are that it increases students' ability to adapt to and participate in change, to deal with problems and make reasoned decisions in unfamiliar situations, to reason critically, creatively and holistically, to collaborate with others and appreciate their points of view, and to identify their own strengths and weaknesses and remedy the latter. It is hoped that the present chapter can offer some insights in relation to these claims.

The modules described in the present chapter have been developed such that they represent a hybrid approach that combines problem-based learning (PBL) and traditional lectures and seminars. Each module Level incorporates progressively more PBL activities, supported by an increasingly sophisticated use of the virtual learning environment WebCT. For example, the Level One and Two modules feature individual PBL sessions that comprise co-operative learning exercises in the lecture theatre paired with Internet based research exercises, while the Level Three module is entirely PBL based. It will be seen below that this Hybrid approach was adopted for three important reasons.

Firstly, the module tutor finds that the Level One students need to be inducted into PBL. Guidance on what is expected of them is important, as it may be quite different from the kinds of learning they have experienced previously, and indeed in other modules during their present studies. For example, the students initially find it quite frustrating that there is no right answer to a problem. Secondly, the tutor has reservations about requiring students to engage in self-directed study with no initial background information, as in the 'medical school' model of PBL. The latter is inadequate if evaluated against Vygotsky's (1930) social constructivist model of learning, which holds that learning is mediated by the interplay between the learner's level of cognitive development and their social milieu. Hedegaard (1996) outlines an example of teaching that is based upon Vygotsky's (1930) principles, where tutors provide 'scaffolding' to help children to extend their everyday concepts to include theoretical concepts. In the three modules described in this chapter, the content and workshop activities are structured to initially activate the students' prior knowledge, and then introduce sub-problems in a logical manner so as to extend and build upon these foundations. The modules are designed so that the Level Two and Three modules build upon discipline-specific skills and knowledge that the students have acquired in the previous Levels. For example, Level One requires no previous knowledge of earth sciences. Basic processes, theories and principles are introduced, and case studies are used to illustrate the distribution and magnitude of natural hazards and factors that control the intensity of natural disasters. Finally, the adoption of a full PBL approach would have implications for the nature of the assessment activities on the modules. The tutor didn't want the entire coursework component to be group work, as the comments of students on other modules convey a feeling that the latter can reflect unfairly on 'drivers and passengers.' Their concerns become more vociferous as the proportion of marks allocated to a group project increases.

How the Communication & Information Technologies Have Been Integrated

It will be seen that in the Natural Hazards modules described in this chapter, the VLE is used in a manner that supports the changing nature of students' learning as they progress through Levels One, Two and Three of their studies. The module tutor finds that the students need to be inducted into using WebCT for two reasons. Firstly, few have any experience of using a VLE. Secondly, few other lecturers in Disaster management currently use WebCT, possibly because they perceived to be user-unfriendly, possibly because of concerns that posting materials on the VLE might discourage students from attending face-to-face sessions. In consequence the students are not in the habit of using the VLE, and their progress beyond being 'novice users' is slow.

The deployment of a VLE in teaching and learning on the module does not have an adverse influence on the development of students' interpersonal and communication skills. The PBL aspect of the modules combines with the VLE to produce a scenario where groups of students work round a computer terminal and use such skills during their discussions. On the Level One module, students participate from the outset in a number of residential simulated emergency field exercises, as group and leadership skills form an important part of the programmes. This combines with the particular nature of the students who are attracted to disaster management, such that the cohort bond and mature perhaps more quickly than those on other programmes. As a result, the students tend to work together effectively during short group work activities in the classroom, focussing on the academic content of the problem. The students use such skills continually throughout their three-year degree programme, which contributes to their ongoing development.

How the VLE is used at Level One

The Level One module incorporates PBL to a limited extent, as the students must acclimatise to it and the student cohort is too large to be accommodated in the existing computer workshops. As a result the students' work must be directed to appropriate information disseminated during lecturers. WebCT is thus used largely as a passive medium for transferring information from the course tutor to the students, although tutors also direct their individual work through its 'module mail' and the 'discussion forum' facilities.

During the lectures, the tutor 'talks around' MS-PowerPoint presentations. The detail on the slides is limited; they contain key diagrams and basic information about the processes under consideration, a case study and maybe 'something extra.' These key points are expanded during class discussions and recorded on a white board. Originally the PowerPoint materials were designed as lecture primers to be read by the students so a basic level of knowledge could be assumed in the lecture, although in the past it was found that only a small number of students did so. However, it is felt important that students have access to copies of the slides to act as memory triggers. No hard copies are provided during lectures; the files are uploaded in to the WebCT 'Contents' page. Since the slides contain only key information, the 'perceived' problem of students' non-attendance at lectures due to their availability online is minimised. Each lecture session is further complimented with a WebCT page featuring extra case study information, links to websites and a few optional short formative assessment exercises.

The lectures are linked to several workshop sessions that are distributed through out the year. Although not true PBL in structure, they act to introduce the greater use of PBL approaches in the subsequent Level Two and Three modules. A scenario is posted on the VLE, including details about the geography and geology of an area. Students are expected to read and make notes from the information prior to the workshop sessions. During the workshop an 'event' occurs, for example an earthquake, and the students must work in groups, initially to identify its effects and define intermediate requirements towards devising a response, and then to devise a response.

Several textbooks are recommended for students at Level One, so they need not rely on materials posted on the VLE. WebCT provides a facility for tracking students' access of its contents pages, and this demonstrates that approximately half of each new cohort of students look at the materials regularly right from the start of the course. The module tutor has observed that approximately a quarter will print off the PowerPoint slides on paper and place them in a folder, slotting handwritten lecture notes in between the relevant print outs. While this represents a somewhat limited use of the VLE, it is effective in enabling a significant quantity of information (pictures and maps etc.) to be made available without the University incurring high printing costs. While these costs are potentially passed on to the students, they do however have the option of 'paperless' working, for example by cutting and pasting relevant parts into a word-processed document.

How the VLE is used at Level Two

The content of Level Two focuses on monitoring and predicting natural hazards and communicating risk to the public. There exists no single suitable textbook for this module, as those available are often written for a different audience, geography or politics students who require a far more detailed knowledge of discipline-specific issues. The wide availability of Internet-based material covering natural disasters is thus a boon. A Disaster Management student can be thought of as a Jack of several trades, and the VLE is a useful tool to transfer key information that has been edited so as to be at the appropriate level. Another difference compared to the Level One module is that it is possible to accommodate the entire student cohort in a computer workshop. These differences, along with the students' acclimatisation to PBL during the Level One module, contribute to the greater emphasis on workshop sessions and use of the VLE compared with the Level One module.

At Level Two, a peer-learning approach is adopted. A key component of PBL is the seminar/tutorial small group learning experience, and this is the basic format of the Level Two workshops The workshops are loosely modelled on the approach outlined by Orsini-Jones and Davidson (1999), where a tutor poses key questions that the students address by summarising key information, and then post their summaries on the forum so they can learn from each other. The VLE facilitates this approach with far larger groups than perhaps might be otherwise be accommodated. Typically the students examine research papers in groups and summarise the key findings -each group having a different paper. Due to time constraints, the outputs from workshops, summaries of the research papers, are posted on the VLE. The students are encouraged read them and question each other after the workshops on the 'discussion forum.' This helps to develop their skills in assessing and reflecting upon their own and others' work, a process guided by the tutor in sessions that follow. However, even though the module description provided to the students states that they should support each two-hour workshop or lecture slot with three hours of independent study their response has been poor, and the discussion forum facility is under-utilised. This raises questions relating to the students' motivation, an issue dealt with in detail in a later section.

How the VLE is used at Level Three

The Level Three module covers methods for managing complex natural hazards and their influence on the degree of economic and social development attained by a country. Again the student cohort is small enough to accommodate in a single computer workshop and no suitable core text exists. Approximately 20 weeks of the 30-week long module is PBL orientated, with a few traditional lectures interspersed to introduce new course content. The structure of the module is driven largely by its assessment, which has two components. The first comprises assessments of the multiple hazards to which a developing small island state is subjected. The course materials were developed in collaboration with the Government of Vanuatu, which has kindly provided real data and documentation covering hazards and development, a rich source of background information for the task, the majority of which is accessed through the VLE.

The second is the development of a risk reduction strategy based on the findings of the previously completed hazard assessments. Each two-hour long PBL session is structured and guided in terms of its topic area, which is derived from those things that must be achieved in order for the students to successfully complete the module assessment. Locations of data and or information are provided on the session's WebCT page, as well as materials such as pens, graph paper, maps and computers. Ideally if solutions are not found by the end of a session, the students should complete them before the following week, however, they tend not to do so. Once again, questions relating to the students' motivation are raised, and they will be dealt with in a later section.

For the hazard assessment component, the students are required to work in groups of around four and to assess a hazard using relevant raw data, and to evaluate the limitations of that data. Spreadsheets of raw data are provided in a MS-Excel file format regarding the recent history of earthquakes, cyclones and volcanic activity in the vicinity of the island. In the first workshop, the students are required to follow a worked example of the earthquake data over a short timeframe. The following 'lecture' introduces a discussion regarding the limitations of using such secondary data. In the second workshop students must analyse the cyclone data. A discussion of the nature of cyclones introduces the problems associated with the fact that data regarding the size and energy of cyclones are rarely recorded in detail; indeed, there may be no data at all for the period when it passes closest to the island. The students must devise a methodology to rank cyclone hazards, for example, they must decide if a large magnitude storm passing the island 200 km away will be more or less damaging than a storm of smaller magnitude passing at 100km. The students are expected to seek all information they need to show they have made an informed decision, such as historical records of damaging cyclones, variations in size of storm etc. There is no single right answer; several different methodologies exist, each of which yields different ranking and therefore recurrence intervals. The results are submitted as a 'website,' and each groups' analysis is displayed anonymously on the VLE.

A seminar focuses on the analysis of the various methodologies that each group of students has deployed, a means of giving feedback to the groups and having them identify errors, misconceptions, and areas for improvement. This approach has proven itself an excellent way of introducing the notion that students should understand someone else's methodology before employing either it or their findings in their own work.

One highly significant benefit is that many students recognise the potential pits into which they may fall -making management decisions on the basis of results and analyses they take for granted to be 'right.'

The second component of the module assessment is an individual task. A static curriculum cannot accommodate all possibilities, and the aim is to build-in flexibility for the students. Each student chooses an area of interest, linking their learning with their own interests and motivations. Each deals with a different hazard in a different geographic location, to prevent them from copying each other's work. This is followed by an 'open book' exam based both on those materials and paper copies of documents containing data on the population and so on of the island that are held in the department. The students must use the information to individually design a 'risk reduction strategy' against a specific hazard for a specific location on the island.

Issues Related to Integrating Communication & Information Technologies

The approaches to the three modules that have been described above raise several issues, which broadly fall under four categories. These are University administration systems, students' Communication & Information Technologies (C&IT) skills, harnessing their accumulated knowledge as they progress through the three modules, and their motivation. They are dealt with in turn below.

Issues Related to University Administration Systems

The students' initial use of the VLE has been up until now significantly affected by administrative issues. The speed at which students are granted access to their modules through the VLE at the start of their course is crucial. For example, in previous academic years, it has been at least four weeks before the entire cohort was able to do so. A delay of this kind makes 'starting as one means to go on' very difficult. The tutor had to provide paper copies of introductory material to be given during lecture time. After four weeks students had begun to expect this, and were perhaps reluctant to begin to access the same material from the VLE. This period is also the crucial time to 'keep' students on the module, when those who are able to choose modules finalise their choices.

Issues Related to Students' Communication & Information Technologies Skills

A key issue that influences the designing of the Level One module is the broad range of C&IT skills possessed by the students when they begin their studies. Due to the maturing UK education system, 18 year olds who are progressing from colleges of further education are often far more computer literate than mature students who have returned to study. Coventry University runs a two-hour induction session and generic self-study guides on how to use WebCT, and for the majority of students across the university, this works effectively. However it is insufficient for a significant number of the mature students who undertake the module. In consequence, a considerable amount of the course tutor's time is spent supporting these students.

Issues Related to Harnessing Students' Accumulated Knowledge As They Progress Through the Modules

At Level Three, the module is designed so as to exploit a principle originally outlined by Vygotsky (1930), that an individual's prior knowledge is an important determinant of the nature and amount of new information that they can process. The initial learning activities undertaken by students draw directly on information gained at Level One, two years earlier. However, it has been observed that many of them struggle to remember the details of the key processes that were covered in the Level One module. This has implications for the future development of the module. One possibility is to make the Level One information available on the Level Three WebCT site for the students to use as reference material. This may make the cues built into the Level Three learning activities for activating prior knowledge more effective. This might in turn help the students to internalise the intended learning outcomes more effectively than would otherwise be possible.

Issues Related to Students' Motivation and Time Management

In the Level Two module, the students must study research papers and construct summaries during the workshop sessions. They must then read others' summaries and ask questions about them. However due to time constrains, the latter took place outside the session, on the WebCT 'Discussion forum.' This approach has not worked as well as expected, with the discussion forum remaining largely bare. It seems even well motivated disaster management students are reluctant to do 'extra' work outside timetabled sessions unless it is assessment driven.

Assessing students' contributions to a discussion forum is problematic. The approach taken by Price and Lapham (2002) was to make students' participation in online discussions worth up to 30% of their final grade. However, the marking does not take the quality of postings into account; to ensure their quality, Price and Lapham rely on providing short period of feedback to students at the start of the course. On the other hand, adjudicating quality is problematic: the peer-to-peer discussions will reflect the process of learning, rather than being a perfected and finalised piece of work. Ways are presently being sought to make the most of the discussion forum facility. Next year it is intended that a second session will follow the first so the tutor can guide the students' analysis of the summaries.

Similarly in the Level Three workshop tasks, if solutions are not found by the end of a session, students should complete them before the following week. The course tutor's experience is that this does not always happen, and a backlog may be generated as a result of the students' poor time management. One solution under consideration is for the students to submit each task weekly, as a 'portfolio' of work. However, this may inhibit students from retrospectively modifying and improving earlier work as a result of their developing skills. It remains unclear whether the students would actually revisit their earlier work in this manner unless some incentive was provided.

Students' Reactions to the Deployment of WebCT

This section explores empirical evidence of the students' use of and feelings about the VLE. WebCT includes a 'Student Tracking' facility, which counts the number of times that each student has accessed a module page through its homepage. In the academic year beginning 2002 for example, 64 students completed and passed the Level One module. The site was accessed on average 154 times per student, just under three times a week. The highest number of times that the module's WebCT site was accessed by an individual student was 455. The lowest was 23, roughly once every week and a quarter.

An emergent issue is the relationship between the module assessments and the students' engagement with course activities. The patterns of the students' use of the VLE across the duration of all three modules shows a peak in the numbers who access course content just prior to the exams. At this time, those students who have not utilised the VLE very much (or indeed attended lecture sessions) engage in a frenzy of activity that is clearly assessment-driven. The flip side to this phenomenon is arguably evident in the Level Two students' uptake of the discussion forum, which has been poor, perhaps because some ways of using the forum are non-assessed and so there is no extrinsic incentive to participate.

Regarding their feelings about the online support, for the past three academic years between 2001 and 2004, across all three Levels the average response across each student cohort has been consistent agreement with the statement "I find the on-line learning support for this module helpful." This evidence of a positive response is corroborated to an extent by the module tutor's anecdotal evidence that many students express their support for the use of the VLE, and frequently call for it to be used more widely by other tutors. The students have also given their opinions of the learning and assessment methods described in the present chapter. Across all three Levels the average response has been agreement with the statement "The teaching, learning and assessment methods are effective."

Turning next to students' attainment, their performance on assessed work across the three modules provides evidence of their intellectual development. The marking scheme for the course specifies that to attain a mark of 60% or above, students must demonstrably apply higher order skills, such as critical analytical skills for example. In the academic year beginning 2002, they comprised 42% of all students at Level One, 67% at Level Two and 59% at Level Three. Focussing on marks for the most C&IT dependent piece of coursework at each Level, 60% of Level One students, 63% at Level Two and 60% at Level Three, which suggests that the use of C&IT may offer particular benefits for students during their initial progression to higher education. Overall, these data suggest that the students' ability to develop a deep understanding keeps pace with the increasing complexity of tasks and skills built into the modules across the three Levels of study. Thus it seems that the combination of a VLE and the hybrid PBL approach described in the present chapter offer an effective means by which to foster the students' intellectual development.

The next consideration is those students at the other end of the scale who achieve marks below 60%. They further sub-divide into a relatively small proportion of students who achieve 50% and below; the assessed work falling into this category demonstrates little in the way of higher order skills. For the academic year that began in 2002 they comprise 6% of all students at Level One, 12% at Level Two and 3% at Level Three. It is arguable that such students adopt broadly a 'reproducing orientation,' to their learning (Entwistle and Ramsden 1983; and Biggs 1987), memorising lecture notes without engaging intellectually with the content. It is unclear whether these figures reflect some aspect of the modules' design or individual students' motivation and study habits.

Issues and Future Plans to Modify or Enhance Technology Integration

Throughout the following discussion it will become clear that the institution-wide availability of WebCT at Coventry University has impacted positively on the implementation of PBL in the modules discussed in this chapter. It will be seen that the VLE has an important role to play in enabling the module tutor to meet several challenges that will need to be faced in the near future.

The first issue to be considered is the diversity of student programmes undertaking the modules. The three modules were originally designed for International Disaster and Engineering Management (IDEM); some time later, Development and Health in Disaster Management (DHDM) took them up. More recently, Emergency and Disaster Management (EDM), which focuses on managing emergencies and disasters, and Environmental Hazards in Disaster Management (ENVDM), which focuses on environmental hazards, have taken up the modules. As might be expected the latter two additions have a slightly different focus from the former. While the modules' primary focus on management is equally relevant to students on all programmes, the question is how to tailor the same modules for the differing focus of each programme. The VLE offers the means by which to make learning resources and other course materials available that are tailored to the needs of the students on each programme. ENVDM students are provided with data relating to, for example, anthropogenic pollution, for example, anthropogenic pollution, faecal colliform in marine water samples and pesticides fertilisers in surface and groundwater. For students on the EDM programme, which has a developed world application, issues relating to disasters in developing countries, while of interest, should perhaps not be a focus. To provide for those students, modifications are presently being made to the Level Three module's WebCT pages to incorporate data and information from the UK Isle of Wight, a small island adjacent to the English mainland.

Another consideration is the impact of making course materials such a lecture handouts available electronically. Some would argue that this might be a disincentive for students to turn up to lectures. However, the tutor of the modules that are discussed in this chapter has found that this is not inevitable. Neither textbooks nor WebCT alone are sufficient for the students to develop the expertise necessary to pass the module. This highlights the tutors' role, which is to 'add value' during the face-to-face aspects of the modules, sharing their expertise and experience with students in a way which is responsive to their particular interests and concerns. Others have come to similar conclusions; for example, Bell (2002) found that posting lecture notes on WebCT for students to discuss among themselves led to increased attendance at his lectures. Bell suspects that this frees the tutor to concentrate on making the lecture experience memorable, and the students find this attractive. Interestingly, a questionnaire survey of 209 students across Coventry University revealed that 4% of the 184 respondents who report that 'a few,' 'most' and 'all' of their modules use the VLE agree with the statement "the notes are on WebCT, I don't go to lectures" (Bromage 2003).

The nature of PBL is such that it requires a large corpus of learning resources. Developing the course learning materials over the past three years has been a labour-intensive process. Even now, they are still perhaps not as comprehensive as they need to be, in terms of providing the guidance and scaffolding to facilitate all students' attainment of the learning outcomes specified in the modules. In the hybrid PBL approach adopted on the modules discussed in this chapter, the problem solving requires that the students' seek a significant portion of the required material from suggested journals, textbooks and Internet sources, whereas in 'true' PBL, case study materials are provided. This has the advantage that each year, the course materials are continually augmented, as the references utilised by previous cohorts are made available to the subsequent year through the VLE.

The earlier discussion of the hybrid PBL approach has highlighted the importance of two key factors for its success or failure. They are students' levels of motivation and their prior learning experiences. Regarding the latter, there exists independent evidence of the impact of students' expectations about teaching and learning upon the success of PBL-based courses. Newman (2003) studied of two groups of nursing students, one drawn across six London hospitals from traditionally taught courses, the other from PBL-based courses. In an article by Utley (2003), Newman reports on his preliminary findings, a significantly higher dropout rate from the latter group, 41% as opposed to 6%. Those dropping out of PBL courses typically make comments that express a lack of clarity about what they were supposed to be doing, and a feeling that no 'teaching' had taken place. Relating this to the modules described in the present chapter, on commencing their higher education, the students tend to have little if any experience of PBL, yet student dropout has not been an issue. Their lack of experience has been taken into account, and the curriculum design such that the PBL component is progressively increased over the Level One, Two and Three modules.

It is arguable that the issue of motivation is related to this in complex ways. Savin-Baden (2000) argues that students who have little prior experience of PBL must learn to recognise and value their own learning experiences and their role in developing both their own learning and independence. Savin Baden (2000) proposes that opportunities for reflection within groups can help in this respect. However, the discussion in the present chapter in relation to the students' motivation to participate in online discussions and workshops suggest that the root of the problem is a 'chicken-and-egg' situation. If the students do not recognise the value of such forums, they may be less likely to participate seriously, and in consequence will have a poor experience. It has been seen that the plan is to encourage their participation by making it count towards their final grades.

In terms of developing the problem-based learning aspect of the three modules, future plans are inspired by workshops and events on problem-based learning and on using C&IT that have been run by the Geography, Earth and Environmental Sciences Learning and Teaching Support Network (GEES LTSN) in the UK. A field trip to Sicily is planned, during which the students will conduct risk assessment exercises. Prior to going to Sicily the students would write a project proposal based upon supplied hazard, population and country profile information. This would be supported by resources set up on the modules' WebCT site, pages introducing the geology of Sicily, maps diagrams and so on, and links to useful external websites. To give the students prior knowledge of the particular sites they would visit, videos would also be loaded onto the VLE. On their arrival in Sicily, the students will spend a day working at a site to collect the information that they had pre-specified in their project proposals.

A current priority is to improve accessibility for students with disabilities. This is of considerable importance to the University, which must comply with recent UK Government legislation that seeks to extend equality of opportunity to those with disabilities (Chapter 10 of the Special Educational Needs and Disability Act (2001), recently introduced as law in the Disability Discrimination Act Part 4). Disaster Management tends to have an image of 'macho types' running around on mountains and surviving in the wild. However, in many ways the role of those who sit behind desks and co-ordinate events is more important. Using a VLE opens the possibility of developing the approach outlined above into a 'virtual field trip' for students with mobility issues that would otherwise prevent them from participating. Such students, as well as those with other disabilities, already benefit from learning materials being made available online through the VLE. Both the existing availability online of learning materials and the proposed virtual field trips offer benefits to other students who might find it hard to attend a field trip, for example, those with caring or daytime work commitments.

Across the UK Higher Education sector, there is an increasing tendency for students to choose their University not vice versa. Furthermore, more and more other universities are developing undergraduate programmes in Disaster Management. Thus in the future it will be necessary to sell the whole package to students. One way to do so is to offer students 'added value': demonstrable advantages when compared to similar programmes at competing Universities. It is also advantageous to attract a wider student intake than 18 year olds who are progressing from further education colleges, particularly mature students. One way to both add value and make learning experiences more attractive to mature students is to employ a wider variety of teaching methods than 'traditional chalk and talk.' It can be seen from the discussion above that the hybrid PBL approach described in the present chapter offers much potential in this respect.

References

Barrows, H.S., and Tamblyn, R.M. (1980). *Problem-based learning, an approach to medical education*. New York: Basic Books.

Bell, S. (2002). *Talking pictures on the web*. Retrieved July 27, 2003, from
http://home.edu.coventry.ac.uk/webctwise/121GD/121gd_ov.htm

Biggs, J. B. (1987). *Student approaches to learning and studying*. Hawthorn, Victoria: Australian Council for Educational Research.

Bromage, A. (2003). *The relationship between students' orientations to learning and their use of and feelings about a VLE*. Paper presented at the '2nd Global Virtual Learning and Higher Education' conference, September 12-13, 2003, Mansfield College, Oxford.

Deepwell, F. H., & Syson, A. J. (1999). Online learning at Coventry University: You can lead a horse to water . . . . *Educational Technology & Society*, *2*(4). Retrieved July 27, 2003, from http://ifets.ieee.org/periodical/vol_4_99/deepwell.html

Entwistle, N. J., and Ramsden, p. (1983). *Understanding student learning*. London: Croom Helm.

Hedegaard, M. (1996). The ZPD as a basis for instruction. in H. Daniels (Ed.) *An introduction to Vygotsky*. London, Routledge.

Orsini-Jones, M., and Davidson, A. (1999, 10 September). From Reflective Learners to Reflective Lecturers through WebCT. *CTI active learning*, (pp. 32-38). Retrieved July 27, 2003, from http://www.ilt.ac.uk/public/cti/ActiveLearning/al10pdf/index.htm

Newman, M. (2003). *A pilot systematic review and meta-analysis on the effectiveness of problem based learning*. Middlesex: Project On The Effectiveness Of Problem Based Learning (PEPBL). Retrieved July 27, 2003, from http://www.hebes.mdx.ac.uk/teaching/Research/PEPBL/index.htm

Price, M.L., and Lapham, A. (2004). The virtual Seminar. In D. Seth Preston (Ed.). *Virtual learning and higher education*, (pp 15-28). Amsterdam: Rodopi.

Savin-Baden, M. (2000). *Problem-based learning in higher education: Untold stories*. Buckingham: The Society for Research into Higher Education and Open University Press.

Special Educational Needs and Disability Act 2001: Chapter 10. London: The Queen's Printer of Acts of Parliament. Retrieved July 27, 2003, from http://www.hmso.gov.uk/acts/acts2001/20010010.htm

Utley, A. (2003, 23 May) Misfit for purpose? *The Times Higher Education Supplement, Pull-out special feature*, pp.12-13.

Vygotsky, L. (1930). *Mind and society*. Cambridge, MA: Harvard University Press.

Integrating Technology in Introductory Earth Science Classes

Neptune Srimal
Florida International University

Abstract

Multimedia class lectures greatly enhance learning in visually oriented undergraduate introductory earth science courses. Activities based on accessible internet resources allow easy introduction of topics outside of text books, including topics of local interest and up-to-date real-world data on long term events like global warming and the hole in the ozone layer. With wider availability of high bandwidth connection and technological breakthroughs in real-time video conferencing and broadcasting, distance learning is poised to lose some of its asynchronous character. Even in a traditional college class, students are overwhelmingly appreciative of integration of technology and web resources in class lectures. The teaching community needs to reinvent itself in order to utilize the full potential of modern technological resources.

Introduction

I teach introductory level Earth Sciences, Oceanography and Environmental Geology Courses at the Biscayne Bay Campus (BBC) of Florida International University (FIU). The Introduction to Earth Science courses contains more than a hundred students and the other two classes contain about two-dozen students each.

Teaching Earth Science

Unlike the other physical sciences, earth science is an extremely field-oriented science. It deals with deserts and glaciers, mountains and oceans, rocks and sediments, ground water and rivers, earthquakes and volcanoes. In each and every class we are discussing places as remote as the Himalayas, or Mt Kilimanjaro or the Antarctica, to name a few. It is a very visual science. No amount of textbook description can express the difference between a slide and a slump, a debris flow and an avalanche, about the ways a volcano erupts and the erosional landforms of glaciers. Geology is also a science that deals with a dynamic earth. Unless the dynamic picture is brought to the students, it is impossible to give a good foundation to the understanding of earth processes. A field picture of erupting Mt St Helens or of the San Francisco earthquake immediately makes a student aware of the influence of earth science on his or her life and establishes a "bond" between the student and the earth and earth processes. No amount of class lectures can establish this bond. On the contrary, verbal descriptions of natural processes tend to make lectures boring, resulting in a mental "clamming up" of students.

Geology is a science that directly influences our lives. It possibly does not matter to a layperson whether the Big Bang occurred 10 billion years ago or 15 billion years ago. However, a layperson is very concerned whether he or she lives in a high earthquake or flood risk zone. S/he would like to know how to determine these risks, how to alleviate the problems and how the earth scientists are tackling these problems. It is very rare for the textbooks to focus on the geology or environment of one's "backyard." For example, Florida has unique geologic and environmental features. Tectonically it was part of Africa and got attached to US during the last continental collision that formed the Appalachian Mountain. It is a limestone country with related problems like that of sinkholes and fast weathering. It is a coastal state with problems of coastal erosion and infiltration of salt water to inland aquifers. It is home to the everglades, a site of major hydrographic modifications (Kissimmee River modification project and now the everglades restoration project). The coastal rocks and sediments record the Ice ages and global warming. Global warming and sea-level rise will greatly affect the Floridian coastline. Unfortunately it is virtually impossible to expose the freshman students to these very vital and interesting aspects of Earth Science within the constraints of the timetable of a semester, the syllabus and class lectures. I often have to make a trade-off between covering the syllabus and focusing on the Geology of Florida.

Problems of Teaching A Large Class

Teaching a large class has its own problems. It is difficult to establish communications with students (barring those who generally sit in the front rows). By the time we get to know the names of our students, the semester is almost over. This kind of anonymity breeds a vicious circle of disinterest and low grades. The lectures are generally tailored for the average students and it is difficult to introduce challenging topics. Every semester I feel very acutely that the students need:

- a daily dose of quiz on the chapter covered
- material on the local geology to reflect on
- an opportunity to take a self-guided tour of the wonders of world geology.

Unfortunately, the race to cover the syllabus and the lack of opportunity to interact with a class of hundred students, keeps these goals unattainable. I have turned to technological solutions to alleviate some of these problems.

Integrating Technology in the Classroom: My Personal Experiences

All my lectures are now presented as multimedia PowerPoint presentations. This has a number of benefits:

a) The material can be presented in a more organized manner. The students can immediately visualize the logical structure of the presentation.
b) PowerPoint presentations make the lectures visually more appealing.
c) The ease of modification allows me to make quick last minute changes, which in turn enable me to incorporate late breaking topical information into the lectures. For example even as the Mars Rover is sending the first pictures from Mars, I have been able to show the Red color of Martian Soil which is because of the presence of Hematite (Iron Oxide), a mineral that had just been introduced to them (http://marsrovers.jpl.nasa.gov). I could show wind faceted rocks on the Martian surface that are called Ventifacts and are also a feature of terrestrial deserts. I could also show the sand dunes on the Martian surface and how the

dunes enable one to find the predominant wind directions on Mars.

d) PowerPoint presents a platform in which text, video, and sounds can be easily integrated.

e) Since Geology is a very visually oriented subject, videos are always a big help. However, most of the educational videos we have are of 20-40 minutes duration so we can afford to show only snippets from such videos. Often, I was reduced to scrambling between the laptop and the VCR and frantically trying to find the particular sections of the tape I wanted to show. That was very inefficient, time consuming and distracting! I have now started capturing small video clips and incorporating those within my PowerPoint presentations. This enables me to capture sections from different videos as well as from TV and seamlessly integrate them all into the PowerPoint presentations. Although I use an old ATI All-in-wonder video card for video capture, more efficient and cheaper devices are available these days.

f) I try to get an Internet enabled classroom for my lectures. Many of the classrooms in one of the university campuses have such facilities now. All one needs is an Ethernet cable to connect the laptop to the network port in the wall. With the internet enabled, I can quickly navigate to sights of geological interest. Fortunately, there are numerous internet sites that contain items of immediate and often real-time geological interest. For example, the Mars Rover site contains daily pictures and results of scientific studies done on Mars. The websites of the US Geological Survey, National Oceanic and Atmospheric Administration (NOAA), Environmental Protection Agency (EPA), South Florida Water Management District (SFWMD) and the websites of many such agencies contain many up-to-date real-time terrestrial and satellite data. To reinforce student's familiarization with internet research, I give several extra credit homework assignments requiring visits to and researches from many of these sites. For example, I give them a quiz that focuses on the Biscayne Aquifer, which is the aquifer that supplies water to Miami and to South Florida. Information about the Biscayne Aquifer is not available in any of the undergraduate textbooks. The students get exposed to local information and data and at the same time can see how their classroom learning is relevant in their everyday lives. My aim is not only to introduce internet based research but also to expose the students to the wealth of data at these fascinating sites which will kindle their curiosity and encourage them to explore the site further. I also provide a number of links to interesting earth science related sites that not only act as the starting points of their research but also provide a gateway to places for them to explore on their own.

g) I am also using the internet to place all these PowerPoint presentations, as well as class syllabi, old tests for practice and quizzes related to every chapter on the web. I encourage students to download and print those lectures before class so that they need to take only marginal notes during the lectures. This saves a lot of class time. The time thus saved enables me to have more class discussions, deliver the lecture at a more leisurely pace, pack more content, stick to the class schedule and have a review session before the exams. The quizzes I put on the net are not for credit but can be graded on the net. That encourages many students to go through the quizzes regularly and keep their knowledge refreshed.

h) As I mentioned earlier, Geology is a very visual science. To that end, I (and almost all earth science teachers) use a lot of slides from our own collections in our lectures. Students often used to complain that it was difficult for them to retain the visual memories of slides shown for only a couple of seconds during the class lectures. Now that the images are available in the internet, it has become possible for them to go and refresh their visual memories of those images.

i) My email address is printed in the syllabus. I encourage students to email me with any questions or concerns they might have. Students have sometimes contacted me via internet messenger services like Yahoo or MSN messenger.

j) After experimenting with placing class resources in WebCT, I opted to place all these in open websites. Although WebCT provides an excellent platform for fully online classes, the extra hassle to learn to log-in and to navigate in WebCT site made many students reluctant to use those resources. Moreover, the necessity to provide constant technical support was distracting from my regular lectures. An open website runs much more smoothly and it is helpful to a larger audience as evident from hits from places as far as Israel, Australia, Europe and Pakistan.

Use of Tablet PC

Lately, I have started using a tablet PC for class lectures. A tablet PC has all the benefits of a laptop with some additional benefits. A tablet PC enables one to write on the computer screen. This is particularly beneficial in a large classroom. In some classrooms the projection screen totally covers the writing surface. In order to use the writing surface, I had to turn off the projector, raise the screen and afterwards had to do everything in reverse to project the image back on the screen. Students at the back of large lecture rooms had difficulty reading my writings. Now, my handwritten notes are also projected and are easily readable. I also do not have to interrupt my lectures and waste time to go back and forth between projecting and writing.

A tablet PC is much lighter and can be comfortably carried around. The battery also has a longer life. A fully charged tablet has enough juice to outlast even the longest class. I don't have to carry the power supply and mess with one more cable connection in the classroom.

Issues With Technology Integration

Many students face problems downloading PowerPoint slides. The instructor has to be patient and ideally demonstrate the process during the first day of the class. The PowerPoint slides should have a simple and light colored background to ensure long life to the ink cartridges and to make the printouts clear and legible.

Successful integration of technology in classrooms requires strong infrastructure support. Because of large size of video files and because of limitation of hard disk space in laptop computers, I have to limit the video clips captured for incorporating within the PowerPoint to shorter duration and often to lower resolution. With larger hard disk and/or external DVD players we can afford to capture longer clips and with higher resolution. However, an external DVD burner would be required to burn the DVDs. With the price of internal DVDs coming down to less than $100 this has become a feasible option.

Since not all faculty members have video capture cards and it is cumbersome to carry videocassettes to the University Technology Services (UTS) lab facilities and capture these there, it might be a good idea to issue video capture devices on short loan to faculty. I foresee that in near future all educational videos will come in digital format and it will be easy to edit, copy and paste snippets of video directly into the presentation.

Some of the benefits of tablet PCs are negated since the tablet PCs have to be tethered to the LCD projector. It is impossible for the instructor to walk around cradling the tablet during the lectures. A Bluetooth equipped tablet and projector would be ideal.

Tablet screens lie flat on the table and this makes for poor visibility of the screen. A sloping lectern would be a better alternative. However, this will possibly not be a major problem in near future. At least one manufacturer has now come out with a model of tablet PC that is viewable from all angles. The instructor has to concentrate on the small tablet screen while writing thus losing eye contact with students. The only way to deal with this situation is to carry on with the lecture while concentrating to write on the screen.

Although wireless networking capability is built into the tablet PCs, many of the classrooms lack wireless hotspots. With the rapid increase in throughput and drop in prices, it should be possible to setup a number of wireless hotspots all over the FIU campuses.

Tablet PCs do not have parallel printer ports, making it difficult to print on older printers. However, since most new printers have USB connections, this might not be sticking point in the future. USB to Parallel port adaptors can also solve this problem.

In order to write on the slides, either we have to choose the pen as a pointer option within the PowerPoint or we have to 'print' the .ppt files to a software package called Windows Journal and then make the presentation through the journal. However, Windows Journal cannot show video clips and nor does it have a slide sorter view, thus limiting its utility. Selecting the pen option within the PowerPoint makes navigation cumbersome, particularly in the tablet mode.

Lastly, although the stylus is well integrated with the tablet PC operations, the touch pad in the tablet PC behaves erratically. The cursor makes unpredictable jumps and often highlights sections of text by itself. The tablet technology still needs some improvement to be an effective classroom tool. But even at the present level, it is a better alternative to laptops.

Thoughts for the Immediate Future

Apart from the standard technological integration outlined above, I wish to dwell briefly on general trends in the integration of the web and computer based resources in the teaching of earth sciences. The general use of computer and web resources can be categorized as follows:

a. Referring students to web sites that expand on the textbook content e.g., websites containing atlases of mineral and rocks, with articles on topics of recent interest like those on methane hydrates as a source of energy, etc.
b. Using websites that contain up-to-date information, like the EPA website (http://www.epa.gov/airtrends/sixpoll.html) with latest data on pollution or the USGS website (http://minerals.usgs.gov) with the latest data on mineral production, consumption and future trends.
c. Using websites that contain real-time data that can be downloaded and plotted as part of class assignments and can be integrated with the class syllabus, like real-time data on earthquakes, sea-surface temperature data or ozone concentration data from the orbiting satellites, and monitoring data from ground based volcanological observatories, etc. (http://k12science.ati.stevens-tech.edu/realtimedatasites.html contains extensive links to sites from which such real-time data can be accessed)
d. Using websites as sources of scientific animations and computer graphics. A good source of many earth science related educational animations is NASA's animated earth server (http://aes.gsfc.nasa.gov). It is an internet-based server that delivers on-demand animated visualizations of earth science processes, events, and phenomena to educational applications. A simple Google search for earth science related animations bring up hundreds of sites with interesting animations of earth science processes.
e. Using online databases like the mineral database (http://www.minsocam.org/MSA/Crystal_Database.html) of the

American Mineralogical Society or petroleum reserve database of USGS for class assignments (U.S. Geological Survey Digital Data Series - DDS-60).

f. In some innovative programs, Geographic Information Systems (GIS) and Global Positioning Systems (GPS) are being integrated in classroom studies. One such innovative program is outlined by Orndorff (2002), in which GIS software was used to estimate the effects of global warming. Students used GIS to calculate the volume of polar ice caps and estimated the rise in global sea levels if those ice caps melted. They then plotted the extent of submergence of coastal localities by plotting the sea level rise on contour maps of coastal regions using GIS. In another innovative use of online resources in the field of earth sciences, Huntoon and Ridky (2002) investigated long term global climate changes by accessing short term (last 250 years) and long term (250,000 to 450,000 years) data on greenhouse gases (http://cdiac.esd.ornl.gov), short and long-term data on oxygen isotopes (http://www.ngdc.noaa.gov), information about population (http://www.census.gov/ipc/www/world.html), and temperature (http://lwf.ncdc.noaa.gov/oa/climate/climateresources.html) from publicly accessible websites. The students plotted the downloaded data and made their own interpretations and composed brief written reports.

g. A good example of integrating GPS in an undergraduate course, is found in a paper by Johnson and Guth (2002). As a part of class activities the students used hand-held GPS' to calculate the volume of sand in the Great Sand dunes national monument. In a place with no geological outcrops, the students used GPS to calculate the length and accuracy and mapped the 100 m dash track in the athletic stadium. Here at FIU, we used a GPS to map plots of land in our Biscayne Bay Campus along with conventional tape and Brunton compass in the introductory earth sciences lab class.

Although I use web based quizzes to help keep students abreast of subject matter, the time has now come and the technology is now available for implementing interactivity even in larger classrooms. It is possible to conduct real-time quizzes and get student responses electronically. We had seen some interesting demonstrations using clickers (also called

Personal Response System or PRS or Classroom Performance System or CPS) whereby students respond to projected quizzes with a handheld remote control device or a clicker. Each clicker is individually identifiable and the grades are automatically credited to the particular student (see http://www.einstructions.com and http://www.gtcocalcomp.com/). We need to experiment with and keep a watchful eye on this technology.

The internet is now an easily accessible storehouse of information. The new breed of students is more computer-savvy than many of their teachers. We must strive to utilize and integrate this very important resource in our classrooms. My gut feeling is that we can do away with textbooks in many of the college courses relying solely on materials available in the internet. Just as we try to keep abreast with developments in our fields through journals and newspapers, we also need to keep abreast of internet resources. Fortunately, modern textbooks contain many useful topical links that then can be incorporated into class assignments.

I had earlier attempted and then rejected posting streaming videos within the class lectures. Even at a low resolution, it was difficult to download these videos using a dial-up connection. With wide availability of broadband connections, it is now more feasible to load the videos. We need to negotiate copyright limitations with the authors of these videos so that we can post snippets on class web sites.

It appears that broadband internet connections have reached a critical mass in the USA. Since the Fall 2003 semester enrollment in online classes have started to match or even exceed enrollment in traditional classes. There is a demand from the students for more online classes. We need to design our lectures to take advantage of online platforms and to prepare ourselves to meet this demand. At the same time the teaching community needs to reinvent itself so as not to render itself obsolete in the face of technology driven mass education.

Perhaps the most compelling argument against online classes is the lack of interactivity and student to teacher interaction. A number of technological projects are now attempting to overcome this limitation (please see www.siliconchalk.com). In near future students might follow their lecture web cast on their hand-held PDA and be in contact with the class through their cell phone/wireless modem, which in many instances are now integrated with their PDAs. It is time to start serious planning for teaching in such a not-too-futuristic environment.

This brings us to a fork in the road of online education. Till now, the main emphasis of online education was on asynchronous communication between the students and the instructor. The students were not bound to any fixed class schedule and hardly ever needed to interact directly with the instructor. The technology is now available for real-time online interaction between students and the instructor and very soon this technology is going to be commonplace through cell phone based steaming videos, and through common chat and video broadcasting programs like Yahoo or MSN messenger. One such web-based interactive online class is being offered in the Tsighua University of Beijing, China. In their 'smart' classroom pictures of distant students are projected on one wall. The professor writes on a digital white board that, along with the audio and video feeds, are transmitted in real-time to the students' computers. The professor can call any student by shining his laser pointer on any of the students' images on the wall. The image changes to live video and the student then answers the professor's questions (Huang, 2004).

The Students' Perspective

a. Students' Responses

How is web support affecting students? There is more student satisfaction as reflected in the comments I see in their evaluations of the course. I often see phrases like "His Website is excellent! All class materials plus quizzes are there and it was helpful," "great Teacher. Love online notes," "Prof Srimal makes the extra effort to put his lectures practice quizzes and past tests online which are very helpful." In personal face-to-face comments also students have mentioned that they find the online materials very helpful and find the PowerPoint presentations visually attractive and logically well structured.

b. Student Performance

Earth science educators in other institutions have also found that providing web-based notes along with computer driven lectures have significantly enhanced exam scores and overall knowledge-gain in introductory earth science classes (Durbin, 2002). However, after the initial jump in class average by 5-8% points after I started putting my lectures on PowerPoint and the material online, the class average has not changed in the last year and a half. This means that we now need to do something more or different or new. The interactive class quizzes I mentioned above seem to be one such avenue of experimentation. However, before we start implementing such new technologies, our classrooms have to be modified and all the bugs need to be ironed out. Moreover, instructors should be given some free time to enable them to prepare materials for the new class formats and to master the new technologies.

c. Attendance

With lectures available on the web, are the students feeling less need to be present in the class? In my experience, the attendance in my classes has not gone down. If anything, it might have improved a little. That might be due to the fact that it is easier and visually more pleasing for the students to follow the lectures that are delivered using multimedia PowerPoint presentations.

Conclusions

- PowerPoint based multimedia presentations in the classroom improve course administration, save time and enhance classroom learning.
- Students appreciate posting of contents and quizzes on the web.
- Class attendance is not affected by providing class lectures and supporting material on the web.
- There is an urgent need for technological support for preparing and delivering digital content from the laptops and for wider availability of wireless hotspots in the classrooms for real-time integration of the internet with class lectures.
- With the growth and easy availability of wireless, video-enabled communication, synchronous online education is now becoming a distinct possibility.
- As this FIU experience has shown, a core group of faculty provided with technology and infrastructure support can act as a lead team for integration of technology in the classroom.

References

Durbin, James A. (2002). The benefits of combining computer technology and traditional methods in large enrollment geoscience classes. *Journal of Geoscience Education*, *50*(1), 56-63.

Huang, George T. (2004). China's clever class room. *Technology Review*, *107*(5), 26.

Huntoon, Jacqueline A., and Ridky, Robert K., 2002, Examining Long-term Global Climate Change on the Web. *Journal of Geoscience Education*, *50*(5), 497-514.

Johnson, Marie C., and Guth, Peter L. (2002). Using GPS to teach more than accurate positions. *Journal of Geoscience Education*, *50*(3), 241-246.

Orndorff, Richard L., 2002, Introducing problem formulation and spatial analysis with an example in global warming and sea level rise. *Journal of Geoscience Education*, *50*(4) 357-362.

USGS World Energy Assessment Team. (2000). *U.S. Geological Survey world petroleum assessment 2000–Description and results*. U.S. Geological Survey Digital Data Series - DDS-60

Addresses of Web Sites Mentioned in the Paper

NASA's animated earth server:
http://aes.gsfc.nasa.gov

Oak Ridge National Laboratory–Data on greenhouse gases:
http://cdiac.esd.ornl.gov

A web site that contains links to many other web sites with real-time data related to air, water, weather, earth science, and oceanography:
http://k12science.ati.stevens-tech.edu/realtimedatasites.html

Historic temperature data:
http://lwf.ncdc.noaa.gov/oa/climate/climateresources.html

Mars Rover Home page:
http://marsrovers.jpl.nasa.gov/home/index.html

U. S. Geological Survey–Mineral production data:
http://minerals.usgs.gov

U. S. Geological Survey digital data series–Publication 60
http://pubs.usgs.gov/dds/dds-60/

U. S. Census Bureau: Population data:
http://www.census.gov/ipc/www/world.html

Website of CPS clickers:
http://www.einstructions.com

Environmental Protection Agency (EPA):
http://www.epa.gov

Environmental Protection Agency–Air pollution data
http://www.epa.gov/airtrends/sixpoll.html

Website of PRS clickers:
http://www.gtcocalcomp.com/

American Mineralogical Society Crystal Structure database:
http://www.minsocam.org/MSA/Crystal_Database.html

National Oceanic and Atmospheric Administration:
http://www.noaa.gov

National Oceanic and Atmospheric Administration–Oxygen isotope data:
http://www.ngdc.noaa.gov

South Florida Water Management District (SFWMD):
http://www.sfwmd.gov

http://www.siliconchalk.com

US Geological Survey:
http://www.usgs.gov

USGS–National Earthquake Information Center:
http://wwwneic.cr.usgs.gov/neis/epic/edic.html

Online Training for English as a Second/Foreign Language Teachers

Pamela L. Anderson-Mejías
The University of Texas–Pan American

Abstract

This chapter presents a case study integrating online instruction into the teacher preparation program for English as a second or foreign language (ESL/EFL) graduate students at the University of Texas–Pan American in the Lower Rio Grande Valley of South Texas. A brief review of the nature of the course in the UTPA program is followed by key points in planning for the online version of this key capstone course, the actual implementation and assignments within the course, student responses to their online experiences and finally, suggestions for others who may be interested in developing a similar online course based upon these experiences.

Introduction

At the University of Texas–Pan American we offer our MA in ESL to local students and, increasingly, to remote students as well. During Spring 2002, I taught our capstone course in the teacher training sequence entirely online. There were several factors which influenced the decision to teach this course online among which were my own previous experience using the WebCT platform for augmenting courses in the MA program and the fact that four students were located remotely in three different areas of Mexico–Montemorelos, Monterrey, and San Luis Potosi. Another five students were located in the general UTPA area and could have come to class meetings; however, the director for the Center for Excellence in Education/Distance Learning offered the opportunity to teach entirely online and all the students were enthusiastic. The third deciding factor was that the course involved a rather limited number of graduate level students all of whom had been exposed to augmented online courses using the WebCT platform.

Following is a description of the nature of the course in our program, the planning for the online version of this key capstone course, the actual implementation and assignments, and then the student responses to the course. Finally, I will present some suggestions for others who may be interested in developing a similar course online based upon our experiences.

Background of the Course

This course, Practicum in ESL, is a capstone course for teachers in the University of Texas–Pan American MA in ESL. The program itself is based upon what has been called the applied science model of teacher training (Wallace 1991) where coursework tends to separate out different areas of study as if they were not interrelated. Students in the MA in ESL program take courses in second language acquisition theory and practice (applied linguistics courses), the structure and nature of English as a language (the theoretical courses), and the teaching/classroom oriented courses. They also write two comprehensive exams which are intended to help them integrate the various coursework into a unified whole. The Practicum course, described here, is intended to prepare teachers to move into a reflective model wherein they continue professional growth beyond the course and program. (Wallace 1991) To that end, we use a number of assignments where the students must teach lessons then reflect upon these and those demonstrated by their peers. Earlier lesson reflection is guided by the instructor while later lesson reflection is independent of instructor guidance. The course requires that participants 1) be teaching in an ESL/EFL environment for a minimum of ten weeks, 2) present their teaching in the traditional class via live demonstration and video of their actual lessons taught within their context, and 3) discuss, critique, and work together to improve their skills in teaching second language English students whether in public or private settings, in an English-using environment or one where English is simply a "subject" studied as a foreign language among others subjects in school.

Preparation for Online Delivery

To prepare the course for online delivery, goals and types of assignments were reviewed by the instructor, and presented to the technical design team. Prior to the beginning of the course, the instructor met with the director of the WebCT platform used at UTPA, the instructional design manager, the graphic/video technician, the WebCT server technician, and the staff assignment coordinator. The first step was to review the traditional course and discuss means to implement similar pedagogical goals via the online medium. We decided to present material in a modular format and divided the course into four main modules, one preparatory module, and one "catch-all" summation module. I presented the course content material in the four modules where module one was the importance of reflection, reflection via journaling and the location for summarizing of each individual's ten week teaching. Module two was the introduction to concepts to be learned through microteaching, the individual teaching demonstrations (including use of the case study technique as introduction to randomly assigned small group discussion), and the associated paper with reflection from the individual. To introduce the concepts, a case study approach was used wherein students were randomly assigned into small groups to respond to questions following self-selected case studies in the textbook *Teacher's Handbook* (Shrum & Glisan, 2000). Module three was the impromptu re-teach discussion and included our online implementation features using quizzes and chats. The fourth module presented action research as a means to reflect on classroom experience (Nunan 1990), gave sample sections of previously conducted action researches, guidelines for implementation and then included a presentation site for each student to upload her or his action research. In addition, the design team suggested adding a fifth module including a variety of possible assignments on teaching issues covered in our face-to-face courses such as classroom discipline, organization of lessons, cultural bias, testing concerns, etc. Students could select various means of addressing these issues after reading a brief review/overview presented by the instructor. Addressing the issues included such assignments as submitting glossary items, creating a webliography with website URLs and a short bibliography of contents, and site-of-the-day submissions.

After discussing this information with the design team, each member clarified exactly what his or her area would need from the instructor and the other members of the team in order to prepare and upload on time. Needs included such information as time frame (when students had to get information to the instructor so she could segment it and get it to the video technician) and turn-around time for uploading information, amount of information expected (such as length of the teaching videos), quality of source and types of student machine requirements, etc. This information was critical—and became more so as the course actually progressed—since uploading streaming media for access and creating timed release of information for some participants, while not for others, required continual coordination among team members, as well as the staff and course participants through the instructor.

At the suggestion of the technical staff, we agreed on the need for some "learning curve" for the online participants. The team decided to try a dry run using older videos of previous teachers as a model after the instructor secured permission to use these teaching videos. The viewing and teacher-led discussions would occur during the first three weeks of the semester course while the participants were setting up their teaching assignments, meeting their ESL or EFL students and generally getting started in their worksites. Such a dry run would accomplish several things including having students realize some of the difficulties of viewing online streaming media from their personal machines at remote sites, preparing the instructor for dividing student-submitted lesson videos into smaller segments, and giving the design team an idea of how to organize so that when actual video data started to arrive, everyone would be ready. This became an additional module–called the preparatory module–where participants viewed three different ten-minute streamed video segments during the early weeks of the class. Following the viewing the instructor conducted an online discussion of what problems were encountered in these videos when seen through streaming media online. The hope was that, in their own videotaping, similar problems such as focusing on a whiteboard that simply showed up as glare, or not using a personal microphone, could be avoided.

Overview of the Online Course

Thus, a variety of technological modes were used. Participants first logged on and viewed the streamed videos then participated in a teacher-led discussion on the board while also setting up their teaching assignments and entering a personal homepage for the other course participants. They then read through case studies from source materials and discussed via bulletin board in randomly assigned small groups how the teacher described in the case could have improved her or his lesson. Optionally this type of assignment could be repeated with either the same or another randomly assigned group for a different case study if the students wished.

The next assignment required participants to videotape themselves teaching in their own classrooms and send that video via FedEx to the instructor. We then prepared the video as a streaming media presentation for the groups to view. In the traditional classroom, participants would then have been asked to volunteer for "impromptu re-teach sessions" based on some portion of the demonstration lesson. As this was impractical for our online students, we decided to use timed, selective release for the streamed media followed by a required chat session within a two-hour period. In that manner participants would not have a long period to think of how they might re-teach a portion of the lesson, and, we hoped, the impromptu discussion via chat would elicit suggestions for improvement of the original lesson. As the instructor, I did not want to be involved in these chats so that a more open discussion (where students did not look for an "expert's" answers) could ensue. I decided to use a participant as moderator who would receive via e-mail some suggested questions and prompts at the same time as the streamed media were available for that session. He or she would then download and print these to use as guides during the chat. Following each chat, all participants would be evaluated by the moderator and the moderator, in turn would be evaluated by all participants via use of the quiz function. These quizzes could then be used to give feedback to the participants. The chats were recorded and the instructor could review the transcripts later. Each student, therefore, was expected to be a moderator for at least one chat session discussing the video teaching demonstration of one peer and to participate within eight other chat sessions.

In addition, to encourage participants to consider their classrooms in a reflective, yet professional, manner, each student was required to conduct an action research following steps outlined by the instructor based on Nunan (1990). Each student was given a presentation area to upload her or his action research report and others were encouraged to discuss the results via the bulletin board function.

The final assignment for all students was a weekly journal following a set of eight questions that were to be answered each week. At the end of ten weeks, the participants were to review their journals and write a summary of their strengths, weaknesses, and what they learned about themselves and their classes over the reflection period. This final assignment was e-attached to the instructor for evaluation, but not shared with colleagues based on a group decision made at the onset of the course.

Finally, for 15% of the final course grade, students were to select from a broad menu of possible online activities and from the additional topics presented for exploration their own assignments. Some chose to do an additional case study; one chose to write an extensive glossary of terms for online listening exercises for EFL students; another did a webliography of sites where non-native English speakers could find pronunciation samples. These were presented to the group in the fifth module area within the course.

Thus the technology used ranged from chat and bulletin board, through presentation online and streaming media, to traditional write and submit a paper. The course was guided by the instructor, but would not have been possible without the team of designers.

Challenges During Implementation

In spite of trying to trouble shoot prior to the course, problems did arise. Participant's videos rarely were clear enough technically to come across clearly through streaming media. What is acceptable when watching a two or three minute news clip becomes intolerable when viewing a half hour lesson on science terms used in English. Unclear sound, indistinct student responses, and difficult-to-see overhead lessons all made viewing participants' demonstrations challenging. After the initial two lessons, I asked the participants to send me copies of their handouts, textbook pages, and any media used in the classroom on overhead projections or the chalkboard. I then uploaded these to be reviewed either separately or after watching the streamed lessons. In this manner, the students did not need to suffer quite so much when viewing the teaching demonstration. It was particularly nice if student's computer capabilities permitted opening multiple windows so that materials could be referred to while watching the demonstration through streaming media. However, not all students' machines had sufficient capability to do this.

Another serious difficulty was the synchronous nature of the chat. As originally envisioned, students would be randomly grouped for chat sessions to discuss the different teaching demonstrations. Each chat session would differ in participant composition. In such a way, each student would work with every other student in varying configurations. However, it was very difficult for students to "meet" at a given time for the chat sessions. Those who could meet earlier in the day tended to have problems accessing their online carriers and would often be "kicked off" by incoming phone calls and other "net challenges." Those who could meet late at night in one time zone might be grouped with others from another zone who had difficulties meeting at that hour and/or had internet challenges. Thus varying the chat groups became impossible and students tended into two static groups. In addition, those students who were second language (L2) learners had difficulty typing in their comments. Often by the time one L2 learner would answer a question or make a comment, the chat had moved on so far that the comment/question made little sense to other participants. After four chat sessions, wherein all but one student had been a moderator, and following numerous complaints especially from the remotely situated students, I proposed to the group (and they decided to accept) the use of discussion board forums rather than the chats. Although this appeared to improve the accessibility issues, the change removed the "immediacy" of the feedback and re-teach elements entirely from the course; but it was deemed worthwhile to keep the frustration level down. It did alleviate some of the issues of chats, but not all problems faced by the participants in their demonstration lessons online.

Some had problems accessing the streaming media videos despite the fact that we kept each segment to approximately ten minutes (after the first one where the problem was discovered). This problem was the nature of streaming media which requires time to load and view. Some access problems occurred during the first participant's lesson which had been divided into two fifteen minute segments. Later lessons were divided further into ten minute segments then eventually into five minute segments, but unfortunately this caused the participants' viewing to be rather fractured and the overall lesson points were more difficult to see. To address the key access problem, the technician continued to shorten segments for uploading. This resulted in one lesson being divided into six different five-minute segments. The personal side of the equation also impacted the streaming video and chats. Newer teachers and those less familiar with computers and the online environment, tended to become extremely frustrated. The access problems were handled by one or two individuals by changing internet access servers and by two others by using their university computers late at night after everyone else had gone home. One non-native speaker simply opted to not write during chats—this improved when the discussion area was used.

Another difficulty for the participants was the desire not to "hurt" anyone's feelings so comments tended to be less than honest and often were unhelpful for improvement in the individual's teaching skills. When some participants did comment unfavorably on a lesson, others would often go overboard to make sure the demonstrator did not take the comments too seriously. This appears to be a factor in the online environment, as I have encountered it both previously and since. Real critique when written down is more difficult to elicit than when spoken in a group setting where everyone can see the facial expressions, hear the pauses, and so forth.

A summary of the key challenges shows that these fall into two categories. First those of accessibility and second those of implementation. Access via various types of internet servers, phone lines, DSL, or a university T3 line made great differences in how streaming media were received and in synchronous activities like chats. Unfortunately, such computer "glitches" that are quite common become overwhelming especially to newer users of the online medium and to more novice teachers. Interestingly, Meskill et al. (2002) report that novice teachers tend to see the computer (and associated machines like printers) as entities to blame—and thereby feel a sense of helplessness when faced with what might otherwise be considered minor irritations. One participant in this case study commented, "This machine just doesn't want to work when I try to get into the teaching demonstrations. I'm so frustrated!" Such responses were similar to those of the novice teachers in Meskill's study.

The second type of challenge to this course can be considered implementation issues. Sound quality of videotapes subsequently uploaded as streaming videos for online use, importance of smaller segmentation of these, and writing potentially hurtful comments rather than speaking these affected the learning goals of the course.

Changes for Future Courses

Nonetheless, student reactions to the course were generally very positive. All participants commented that the course was very practical, they had learned a great deal, and they really enjoyed the online environment despite its challenges. However, two participants–one of whom was the major motivating factor in changing from chat to discussion–complained that the "chemistry" of the group was damaged by the change from chats. "In the chats, we were really connecting. When we changed to the bulletin board discussions, we seemed more distant from one another and we didn't give any useful feedback for our teaching improvements [sic]." The other participant also noted that ". . . in general, nobody made any serious commentary on my teaching; like it was already fine that I know it wasn't. I think [the fact that] the discussions that could be read over and over were the reason rather than chat which we thought wouldn't be seen later." All rated the various modules as successful. One key comment was the lack of use for the textbook that had been assigned. One student stated, "It would be better for online courses not to use a book. It was a waste of money in this one."

Student journals contained much useful commentary on the course as it was developing. Earlier in the course, three students felt completely overwhelmed by the amount of work needed. In part, this could be attributable to the learning curve which occurred when first working in a completely online environment; in part, it could be attributed to the modular structure where many assignments were available simultaneously and students needed to organize carefully to figure out exactly how each was going to complete the key work for the course. Of these three students, two were non-native English speakers. One was in a remote location and the other was a local student who "could have" attended a face-to-face class had it been offered. The other student had recently moved to a new location, set up the internet and household in a foreign environment, and quite reasonably was overwhelmed by cultural changes.

By the completion of the course, all members had formed a learning community that shared information and asked one another questions regarding issues in teaching beyond the information required or even encouraged by the course itself. One teacher commented that, "without the others in this course, I'm not sure I could have stuck it out!" Another in the ninth week journal mentioned how one of the best things in his teaching was being able to discuss problems with his peers online. Since these discussions did not take place in the "teacher's lounge" or another discussion area of the course, I believe that these students were using the e-mail functions either in the course or through other means of communicating. Such networking is vital to continued growth as a professional and one of the unexpected benefits of this online practical experience course.

For the instructor, the course could have been a killer. It required an enormous amount of time to read through discussions, presentations, and papers as well as the chat transcripts and participant and moderator evaluations. Not only this, but it required a great deal of time to view each teaching demonstration videotape, decide where to cut the lessons into segments which could be uploaded as streamed media. Then it was necessary to meet the deadlines for turn around time to get the videotapes to the graphic technician who needed usually two weeks to prepare and upload the streamed segments. This worked out well for the earlier teachers submitting their demonstration videos, but toward the end of the course as participants got behind in their work, videos arrived barely in time for the technician to upload them as streamed media–let alone in time for the instructor's previewing and deciding on where to segment each lesson. This is not to mention responding to e-mail and evaluating or grading work! Without the assistance of a great team, I would have been lost.

Nonetheless, I would use this format for teacher training classes again; however, there are some changes I would incorporate. First, I would use it over a two semester course rather than try to do it all in one semester online. What works in the face to face environment of the classroom video demonstrations and immediate feedback orally, or even in writing, plus the impromptu re-teaches and reflection on what worked and why, takes immeasurably longer when it must be uploaded to the computer and sent through the internet. Second, I would want to reconsider means to implement impromptu re-teaches. The immediacy of the re-teaches was lost in the online environment. Since some of the best teaching suggestions I have ever encountered in the twenty-five years that I have worked with teacher preparation have come from these impromptu re-teaches for English language learners, I would like to find a way to accomplish this in the online environment. This type of creative thinking "on the teacher's feet" cannot be easily accomplished over the internet perhaps right now, but as technological advances are made, it may become more feasible.

Lastly, I would not be in such a hurry to change the initially planned teaching format despite frustrations and complaints. As a teacher, we are consistently focused on meeting student needs. As a trainer of teachers, I find that I want to exemplify best practice not only in words, but also in how I structure and facilitate the course. In this case, I believe that did not work. The change from chat format to discussion format, despite the fact that the technical team and I had carefully planned the course to include this feature, irretrievably changed attainment of a key goal—improvement in teaching skills. Of course hindsight is 20/20, but the two students who were most frustrated and dissatisfied with the chats during the early part of the course were both newer teachers and relative novices in use of the computer/internet. I not only did not recognize this factor, but also was too hasty in making the change—it would have been better to simply complete the course and consider the change in a future one. In the name of flexibility, serious critique of one another's teaching skills was abandoned. Even the most frustrated, complaining student recognized that changing from the chat to the discussion led to less serious commentary on teaching practices.

Suggestions for Developing A Similar Course

For those considering online courses for teacher training, I suggest first and foremost not to hesitate. Teacher training via online environments can work and work well, but do be aware of the need for 1) very clearly defined course goals which must translate into 2) very well organized assignments that are 3) timed precisely with little or no leeway for "late work." In addition, 4) do not be overly hasty to change format until after thorough analysis of the potential results. Finally 5) a team approach is vital. It would be virtually impossible for one person to create, facilitate, and evaluate participant work while also having to handle the technical side of streaming media, uploading quizzes, and formatting a platform. In this sense a prepared platform like WebCT is indispensable, but so is the assistance of a team. The professor who is accustomed to being a one-person show must retrain her/himself or be doomed to permanent attachment to the computer! In addition, the instructor of the course must see the computer as a great tool but cannot believe that she or he must know everything about it before beginning. I believe a good sense of humor and ability to be flexible is important, too.

From this review of our course, which is to be replicated again at UTPA when the student need arises, I hope that the feasibility of teacher training via online delivery can be seen as a clearly possible task. This case study report delineates a type of online training that is increasingly needed today. Those current teachers who wish to retrain or continue their education can, through the online courses, continue working full time and not need to leave their family and support system in order to get their education. In a field such as English as a second or foreign language, where many teachers live and work in other countries than those where degrees are readily available, this should prove a revolution. These teachers can continue to develop or can actually get a degree which will improve their social standing in their profession while living and working in their homelands. The culture shock of relocation, often without their families, for a year or two-year-long program can be avoided. Not only does this improve their own success, but it also permits their family to avoid the conflicts that arise when a parent or provider must stop working and move to a distant location to study. In short, the successful offering of teacher training courses via the world wide web, should be seen as a milestone in distance education.

References

Meskill, C., Mossop, J., DiAngelo, S., Pasquale, R. (2002). Expert and novice teachers talking technology: Precepts, concepts, and misconcepts. *Language Learning & Technology*, *6*, 46-57. Last retrieved June 7, 2004 from http://llt.msu.edu/vol6num3/meskill/degault.html.

Nunan, D. (1990). Action research in the language classroom. In J. Richards & D. Nunan (Eds.), *Second language teacher education* (pp. 62-81). Cambridge, England: Cambridge University Press.

Shrum, J. & Glisan, E. (2000). *Teacher's handbook: Contextualized language instruction* (2nd ed.). Boston, MA: Heinle & Heinle.

Wallace, M. J. (1991). *Training foreign language teachers: A reflective approach.* New York: Cambridge University Press.

Integrating the Web Into Environmental Decision Making Roleplay-Simulations

Denise Kirkpatrick
Monash University

Robert McLaughlan
University of Technology Sydney

Abstract

This chapter describes the integration of the web into a roleplay-simulation designed to develop students' environmental decision-making skills. The roleplay was adapted for the web and introduced as a shared learning activity (e-Sim) into undergraduate courses at three Australian universities. Web technologies created a rich 'real-world' learning environment, supported distributed interactions and communications, and supported collaborative learning while developing students' information technology skills and disciplinary knowledge.

What Are Roleplay-Simulations?

The use of roleplays within a simulation framework allows participants to adopt a role that encompasses a set of interests, values and knowledge. They require participants to adopt the role of key personae and behave in role-appropriate ways in response to major events and actions of other personae (Errington, 1997; Feinman, 1995; and Gredler, 1992). In educational contexts, roleplay-simulations can be used to create opportunities in which students can experience the complexities of authentic situations and develop associated knowledge, skills and understandings (Galea, 2001). Mekong e-Sim was based on previous e-Sim designs (McLaughlan & Kirkpatrick, 1999; McLaughlan & Kirkpatrick, 2001; and Vincent & Shepherd, 1998).

Why We Chose the Web?

We saw that using the web could assist us in responding to two major challenges:

1. How to support interaction between distributed students?
2. How to create a rich and realistic, simulated 'world' in which students could collaborate?

The web provides an environment that supports distributed and extended interaction (Hiltz & Wellman, 1997; and Heeren & Lewis, 1997) that allowed us to create a learning environment in which students studying at different universities in different parts of Australia could participate. This allowed the development and operation of a roleplay-simulation that more closely resembled real events and situations because we could run it over an extended time frame.

The resources of the web supported the development of a rich learning environment that included a range of resources, accessible to all students. Integrating web technology into a roleplay-simulation also provides an opportunity for students to develop ICT skills in realistic task situations and acknowledges the ubiquitousness of the web (McLaughlan & Kirkpatrick, 1998; Naidu, Ip & Linser, 2000; Poustie, 2001; and Starkey & Blake, 2001).

The learning management system Blackboard was used to support the roleplay-simulation. Blackboard supported synchronous and asynchronous communication using email, text-based synchronous chat, discussion forums (bulletin boards) and areas for group work.

What We Did

Mekong e-Sim (the online roleplay-simulation that is the subject of this chapter) was developed as a module that formed a central part of different courses in three university programs: Technology Assessment, at the University of Technology, Sydney, Asia-Pacific Development, at the University of Sydney, and Environmental Engineering II, at Adelaide University (with Sepang Institute of Technology, Malaysia).

The Mekong e-Sim is based on a realistic international resource management problem that develops specific knowledge and generic skills of students from different disciplinary and professional backgrounds. The specific learning objectives of the Mekong e-Sim were for students to:

- Identify the political, social, economic and scientific dimensions of decision making in the context of natural resource management conflicts
- Identify the responsibilities and appropriate responses for characters in the roleplay-simulation
- Develop communication, research, critical thinking, negotiation and decision-making skills
- Develop an appreciation of cultural differences and approaches and
- Utilise Information Communication Technology (ICT) skills.

How It Worked

Blackboard was used as a common learning management system across all universities. As previously mentioned the e-Sim activity constituted part of the course that students were studying. Students' instructors introduced the idea of the e-Sim and demonstrated the learning management system, and an online training game was used to familiarize students with using the tools available in Blackboard and to introduce them to other students in the e-Sim.

The Mekong e-Sim activity comprises four key stages.

- During the briefing participants are familiarised with the operating rules of the e-Sim, the geographical context of the Mekong region, and evaluation and participation requirements. Students adopt a persona whose role they will play during the course of the e-Sim (in this case four students collaborate to share a persona). Using a range of information sources students develop an understanding of the responsibilities, views and strategies, of their adopted persona identity that is posted to the web. This guides their actions throughout the events of the e-Sim and provides a source of information to other personae about the characters and roles.
- Once the personae identities are established events of the e-Sim unfold. The interaction phase comprises interactions between different personae in response to events that have occurred and the actions of other personae. Key events are released into the e-

Sim by the controller (instructor). These are modeled on real events such as the announcement of a public hearing on a specific topic or the discovery or restriction of a scarce resource shared by various personae. The events releases cause personae to respond and create new events. Participants are therefore required to operationalise the understanding of their persona and the simulated environment gained during the briefing stage. This understanding is reshaped as participants experience consequences that follow from their actions. Most of these interactions are conducted via email but are supported by private, face-to-face and other synchronous communications between the four members of each persona.

- Online public forums simulate public inquiries into key aspects of the environmental situation that is the focus of the e-Sim. Each forum or inquiry has specific terms of reference that elicit formal submissions and responses from various personae. Personae are also expected to make submissions in response to positions submitted by other personae. The topics chosen for the forums involved a proposal to build a large dam, the development of a natural resource management plan for a lake and the prioritization of alternate Regional Transport Initiatives. Another forum was set up to solicit funding proposals from personae relating to development projects they wished to pursue. During the forums students gain an understanding of the multiple perspectives relating to the topic being debated. The debates in the forums take place over a 48 hour period simulating face-to-face public forums used in the real-world situation.
- The debriefing provides an opportunity for participants to identify what they have learned as a consequence of participating in the e-Sim. Individual participants publish online a reflective account of an incident that occurred during the e-Sim and their consequent learning (via a critical incident report). This is supported by a face-to-face session that uses structured, guided recall, reflection and analysis of the activity based on the experiences of the participants and their understandings derived from producing the critical incident reports.

The four key stages of Mekong e-Sim were designed to support particular types of student interactions, collaborative interaction with each other and active engagement with the content of the module. All learning activities in the e-Sim were designed in accordance with the principles of collaborative learning (Johnson & Johnson, 1991; and Abrami et al., 1995). The e-Sim contained a number of assessed learning activities that contributed to students' final grade in the course. These activities were designed to focus and direct students' learning and to provide an indication of the nature and extent of their learning.

Challenges and Issues

The use of a web-based roleplay simulation or e-Sim was accompanied by a number of challenges. These included issues associated with using one learning management system hosted by one university across three institutions.

The planning of the e-Sim in this distributed, collaborative context required substantial organization and communication to ensure that students received consistent instructions and it was critical that instructors spent time clarifying their expectations of student learning outcomes, engagement and activities. This required some adjustments to course outlines and assessment weightings. The level of student support provided at each site varied and at times this needed resolution. There was variation in the level of student ICT skills that affected individual student's capacity to engage and capitalize on the benefits of the internet.

Because the e-Sim operated in three different universities there was considerable variation in the level of student access to regular and reliable internet access. A significant issue related to staff or faculty workload. The e-Sim allowed students to engage with the activities and each other 24/7. It supported a continuous student experience and this was accompanied by increased student expectations about staff availability. There were also issues relating to student workload resulting from sustained student engagement with interactions within the e-Sim and conflict with other course workload requirements.

While the e-Sim design includes introductory activities that explicitly facilitate the establishment of relationships among group members, this still presents challenges for students who must establish effective working relationships with their peers very quickly.

Addressing the Issues

We have redesigned some aspects of the e-Sim to minimize staff workload in relation to managing students and groups and progressing the events of the e-Sim. For example, our first version of Mekong e-Sim included a large number of cross-institutional personae. These proved to be very difficult and time-consuming for staff to manage and support. In order to reduce demands on staff in supporting students working across institutions we removed cross-institutional personae, basing all members of individual personae in one institution. Peer assessment was also introduced, contributing to the assessment of student participation, and this reduced associated staff workload.

Assessment tasks were redesigned to reduce staff and student workloads while focusing on desired student learning. We reduced the number of summative assessment tasks because student feedback and staff observation indicated that students were focusing on the assessment tasks at the expense of engaging in the interaction of the roleplay. We also substituted online quizzes for an existing essay that was intended to elicit students' understanding of the roles and responsibilities of personae and background to the Mekong region. The quizzes provide immediate feedback to students and staff and also reduced staff marking time, thus freeing them to contribute to the e-Sim. In response to student feedback we replaced a combination of individual and group submissions relating to the final report with a single group submission accompanied by peer assessment of contribution to the persona (Freeman & McKenzie, 2001).

We also developed and refined existing communication and working protocols. For example, we developed protocols relating to 'shared personae' to share student responsibility for accessing the web and posting official communications to the e-Sim site.

The redesign of some activities within the e-Sim and the formation of single-site persona groups have contributed to a more manageable activity and one that allows students to engage more conveniently. The management and operation of the groups sharing personae has been refined to ensure that students receive consistent messages and receive necessary levels of support.

Our experiences illustrate the importance of providing participants with a variety of communications tools and media. Previous work has identified the appropriateness of different tools in supporting different task completion and achieving different outcomes.

Student Responses

The effectiveness of Mekong e-Sim was assessed using a multifaceted approach including the use of student feedback, analysis of student performance on formal assessment tasks, analysis of interactions and staff feedback (Kirkpatrick & McLaughlan, 2001). Generally the students have responded positively to their experiences with Mekong e-Sim. While acknowledging the workload of the activity and some of the limitations presented by inadequate internet access, students perceived that the exercise was valuable for developing 'soft skills' as well as disciplinary knowledge. They believed that this activity contributed to their developing an understanding of the complexities of working with others on a common task (particularly those from other professions or disciplines), and of some of the realities of the real world. In addition, they believed that Mekong e-Sim contributed to their subject knowledge, a perception confirmed by their performance on assessed learning tasks (McLaughlan & Kirkpatrick [in press]).

Conclusions

The Mekong e-Sim extended the notion of single site e-Sims to include multiple sites and multiple discipline involvement supported by the web. The e-Sim effectively engaged students in a collaborative learning activity that developed their understanding of the issues and complexities associated with environmental decision-making. The use of the web enabled staff and student collaboration, and provided the tools necessary for communication and task completion. Mekong e-Sim assisted students in achieving the required learning outcomes and we believe that it provides an example of the successful integration of technology into teaching at the university level to create learning opportunities not previously possible.

References

Abramai, P. C., Chambers, B., Poulsen, C., de Simone, C., d'Appolonia, S., and Howden, J. (1995). *Classroom connections: Understanding and using cooperative learning.* Toronto: Harcourt Brace & Co.

Errington, E. (1997). *Role play.* HERDSA Green Guide No 21. Canberra: HERDSA.

Feinman, J. (1995). Simulations: An Introduction. *Journal of Legal Education, 45*(4), 469-479.

Freeman, M., and McKenzie, J. (2001). Aligning peer assessment with peer learning for large classes: the case for an online self and peer assessment system, In *Peer learning in higher education: Learning from and with each other*, D. Boud, R. Cohen, and J. Sampson (Eds.), pp. 156-169, London: Kogan Page.

Galea, C. (2001). Experiential simulations: Using web-enhanced role-plays to teach applied business management. *Information Technology and Management, 2*, 473-489.

Gredler, M. (1992). *Designing and evaluating games and simulations: A process approach.* Houston: Gulf Publishing.

Heeren, E., and Lewis, R. (1997). Selecting communication for distributed communities. *Journal of Computer Assisted Learning, 13*, 85-98.

Hiltz, S.R. and Wellman, B. (1997). Asynchronous learning networks as a virtual classroom. *Communications of the ACM, 40*(2), 44-49.

Johnson, D. W., and Johnson, F. P. (1991). *Joining together: Group theory and group skills.* Englewood Cliffs, New Jersey: Prentice-Hall.

Kirkpatrick, D., and McLaughlan, R. (2001). Evaluating Pollutsim: Computer Supported Roleplay-Simulation. *Academic Exchange Quarterly, 5*(4), 99 - 106.

McLaughlan, R., and Kirkpatrick, D. (1998). *Developing a framework for effective electronically supported simulations.* Paper presented to the annual Higher Education Research Development Society of Australasia conference, Auckland New Zealand, July.

McLaughlan, R., and Kirkpatrick, D. (1999). A decision making simulation using computer mediated communication. *Australian Journal of Educational Technology, 15*(3), 242-256. http://www.ascilite.org.au/ajet/ajet15/mclaughlan.html Accessed April 28 2004.

McLaughlan, R.G., and Kirkpatrick, D., 2001, Peer Learning Using Computer Supported Roleplay-simulations, In *Peer learning in higher education: Learning from and with each other*, D. Boud, R. Cohen, and J. Sampson (Eds.), pp.141-155, London: Kogan Page.

McLaughlan, R., and Kirkpatrick, D. (in press). Online roleplay: Design for active learning. *European Journal of Engineering Education.*

Naidu, S., Ip, A., and Linser, R., (2000). Dynamic goal-based role-play simulation on the web: A case study. *Educational Technology & Society*, *3*(3), 190-202.

Poustie, M.R. (2001). Engaging students and enhancing skills: Lessons from the development of a web-supported international environmental law conference simulation. *International Review of Law Computers & Technology*, *15*(3), 331-344.

Starkey, B.A., and Blake, E.L. (2001). Simulations in international relations education. *Simulation & Gaming*, *32*(4), 537-551.

Vincent, A., and Shepherd, J. (1998). Experiences in teaching middle east politics via internet-based role-play simulations. *Journal of Interactive Media in Education*, *98*(11). http://www-jime.open.ac.uk/98/11/vincent-98-11-paper.html Accessed April 28 2004.

Using Geographic Information Systems (GIS) Software for Teaching and Learning

Jennifer Gebelein
Florida International University

Abstract

This chapter introduces GIS software as a means of teaching and analyzing spatial data. It makes a case for active learning and active teaching practices that present potentially more substantive techniques toward student comprehension of spatial information in a variety of contexts. These hands-on techniques promote discovery of diverse patterns in complex qualitative and quantitative data. The active learning principles discussed in this chapter have been shown to increase a student's visual learning capacity, which in turn enhances student literacy through map interpretation and creation. The author hopes that more instructors will utilize these techniques for analyzing spatial and temporal data relationships in their classrooms to assist in the understanding of complicated associations among diverse factors.

Introduction

Geographic Information Systems (GIS) provide students and teachers the tools to analyze both simple and complex spatial relationships among the various components at different locations. The importance of utilizing this type of tool is to teach students to look critically at how things change over space and time. The advantage of using GIS to teach this critical thinking is seen in the final product of any GIS analysis. Users of GIS can query the various components in their dataset and the final answer that they are given is in the form of a map. Maps can contain large amounts of information and are an easy-to-understand medium to show spatial relationships that are not necessarily apparent when simply discussed verbally or in essay format. In this chapter, a thorough discussion of utilizing GIS into the classroom is presented, followed by practical and realistic ideas to those who wish to integrate this type of technology into their (non-GIS class) teaching. The sample course that is discussed here is *Introduction to Geographic Information Systems (GIS)*.

Description of the Course

Introduction to GIS is taught at the Florida International University as one of the basic requirements for a Bachelor's degree in Geography. Geography, when literally translated, means *description of the Earth*, but takes into account many aspects of almost every science. The students work with several different kinds of both physical and social science data throughout the class. Physical science data may include satellite data, aerial photography, digital topographic maps, or classified land cover information. Social science data may include elements of economics, ethnicity, population, or taxes. In this course, students learn the basics of data acquisition, data importation, data manipulation, data representation, errors commonly found in data, spatial analysis, and fundamental cartography skills.

First, students are provided a comprehensive historical overview of spatial data analysis and the evolution of both the ideas and concepts behind mapping and the creation of GIS. Then the basic skills necessary to complete simple spatial analyses utilizing ArcGIS[1] software are taught. The class is taught in a computer laboratory. Therefore, almost every class has some hands-on component that takes the lecture and reading materials, and then applies them to the data and some aspect of learning the software. Students are required to complete an independent project using basic and advanced ArcGIS tools. The class project that is due at the end of the semester is individualized based on the interests of each student. The professor and teaching assistant work with each student on an individual basis. Such individual attention is necessary to train the students to use the appropriate advanced spatial analysis tools in order to answer their spatial research questions. Having the students take an active role in their own learning process is the exciting part of this course.

Background Information About Geographic Information Systems (GIS)

GIS is commonly defined as "An information system that manages, manipulates, analyzes and presents spatial data" (Theobald, 2001, p. 10).

Spatial data is literally almost anything that has a location attached to it. For example, a person standing in front of a library holding a Global Positioning System (GPS) device can obtain her latitude and longitude. She is now a point in space, and therefore spatial data. A house has a street address. The address of the house makes the house spatial data. Many cities, streets, major lakes, and ponds have been named and digitally catalogued so they can be viewed in a computer environment. Census data is another very popular type of spatial information. As long as there is a location linked to the feature of interest, it can be called spatial data.

GIS data is arranged like layers on a cake. The bottom layer is called a *base layer* and resides on the bottom of the digital display. The addition of each consecutive data layer forms the basis for any question students might have. A layer may also be called a *theme* because each data layer could have a different focus such as population data, cities data, or political boundary data. Each time a student adds a layer or theme, she is overlaying the base layer with additional data. The layers are displayed by name in a legend alongside the map so it is easy for students to keep track of the data layers they have added. Students may also remove one or more layers from their data layout if they find them unnecessary. Once all layers have been added, analyses can then begin to answer a spatial research question. The spatial research question usually addresses three or more data layers and can be answered utilizing GIS tools. The basic tools in the software provide a means to look for patterns, trends, similarities, differences or anomalies among the different datasets supporting each data layer displayed. In any course that uses spatial information of any kind, GIS can enable students to gain more familiarity with the processes of basic statistics, creating maps, and analyzing data.

How Maps Work and Are Perceived By Students

In a GIS environment, the concept of cartography as a graphic communication is the basis for mapmaking. Cartography can be defined as the art or science of making maps. The art behind any cartographic GIS map, this course, lies in the hands of the students. A map can be made to display the student's queries, agenda or data model. Color choice, symbolization and map layout are all important factors in the successful, or unsuccessful transmission of information from map creator to the reader of the map. If the author's purpose of a map is clear, then the map is successful. If the information received by a viewer is muddled, unclear and lacks purpose, then the map has failed in its attempt at data transmission (MacEachren, 1995).

Map perception involves both how we process certain visual stimuli as well as certain imagery. The production of a functional map involves understanding how, through our vision our brain extracts meaning from maps. A functional map mirrors a successful map in that the purpose of the author is clear to a viewer, regardless of experience. Perception of a map's contents, graphics, symbols and layout can also be heavily dependent upon the viewer's own life experiences. However, with a basic background in symbol, color, pattern and object choice, which can be learned in a basic GIS class most maps can be made accessible to any audience (MacEachren, 1995).

How GIS Is Being Integrated

As mentioned previously, the class is tailored to each individual student's interests. After a general history of spatial data analysis is completed in addition to an introduction to the software, the students are expected to construct a project that is focused around one or more spatial research questions in which they are interested. For example, one student may be interested in whether traffic congestion could be lessened in certain areas with the addition of side roads or overpasses. Another student may be concerned with the current protected areas in Florida and try to find those areas most likely in need of further protection from an adjacent growing urban population. Once a spatial research question is established, with the aid of the professor, students locate appropriate data, integrate that data and begin using the GIS tools to map the answers.

Learning Principles

The purposes of the major learning techniques presented here are to make students learn *actively* and increase their visual learning capacity. Simply lecturing to a class with no or little teacher/student interaction is an example of a passive learning technique. It is important to avoid this type of teacher-centered and student-passive instruction because students do not absorb the material as well as when they are actively learning (Creed, 1997). The purpose of active learning is to prevent a passive learning mode by including activities that engage students in active contemplation and task completion (Mazzur, 1997). One model of active learning includes visual learning. The term "visual learning" was first coined by John Debes (1969). Debes postulated that visual cues supplementing lecture information actually increased a student's ability to retain and understand the information presented (Fransecky & Debes, 1972).

One of the assumptions of visual learning is that all images convey meaning, and literacy means being able to study and comprehend that meaning. The interpretation of images is fundamental to normal human learning. Once established, this interpretative skill enables a visually literate person to distinguish and understand the visible actions, objects, and/or symbols, natural or man-made, which are in a myriad of different environments (Arnheim 1969, Ausburn 1978, Curtiss 1987, MacEachren 1995). This concept of visual learning is directly applicable to the learning that takes place in a GIS class. In fact, GIS is not only a type of visual learning, it is an empowering learning tool because of the ability students have to determine the manner in which data is organized, displayed, queried and mapped. The students, in essence, are creating a highly technical learning environment in which they tailor the information to best express their comprehension of the course content and data.

Achieving Active Learning in the Classroom

Active learning incorporates teaching techniques where students do not merely listen to a lecture. Students are actually engaged in some activity that forces them to mentally manage, interpret, and harness some information to solve a problem or analyze a situation. According to Meyers and Jones (1993), the concept and practice of active learning draws from two fundamental theories: (1) that learning is an innate active exercise and (2) that different people learn in different ways (Meyers and Jones, 1993). Studies reveal that learning that is more retentive occurs while students are actively learning. Lectures are important, and when combined with active learning tactics, can be a valuable part of a student's education. The four essential building blocks of active learning are talking and listening, writing, reading, and reflecting (Meyers and Jones, 1993). Since students learn differently, it is critical that an instructor has several "plans of action" for incorporating elements of each building block into weekly lesson plans. Different teaching techniques work well with some groups of students and not others. Thus, it is helpful for the instructor to have a number of teaching techniques to choose from for any given class.

There are many suggested techniques in the literature for incorporating active learning into the classroom. Presented here are several proven techniques that may be useful for a GIS instructor. According to David Kolb, Professor of Organizational Behavior in the Weatheread School of Management, experiential learning may be one of the most successful learning tactics. Dr Kolb created an active learning model that incorporates concrete experience, observation and reflection, the formation of abstract concepts and testing in new situations. This type of active learning can be achieved by several means including field experiences, computer-based hands-on exercises, development of class projects, or internships where on-the-job experience helps mentally cement in class learning (Smith 2001, Borzak 1981). A second popular and successful approach involves cooperative learning in both small and large groups. Cooperative learning forces students to deal with information in several ways that ultimately and actively, help them learn.

Students listen, read, reflect and then discuss the information with their peers to yield a final project, presentation or some other type of collaborative graded work. Doing this allows students to refine their understanding by talking about their ideas with other students, become aware of other students' perspectives and to achieve group communication skills (Harmin 1994, Silberman 1996, Meyers et al. 1993). A final successful active learning strategy is presentation learning. Activities could include final project presentations, peer teaching or debate. Final project presentations require the student (or students if a group project) to synthesize an entire semester's worth of work into approximately fifteen minutes. This type of presentation is usually done in PowerPoint. Generally they are graded on substance, organization, explanation of technology and map choice. Peer teaching is an excellent and effective way to actively get students involved in the subject matter. Peer teaching can be accomplished through article assignment and student-led discussions each week (Haller et al 2000). Peer teaching can also be accomplished by calling students randomly to the front of the class to demonstrate a GIS skill. Debate is a third successful method of active learning. In a GIS class setting, I often give students opposing sides to the issue of privacy and spatial information. This debate requires them to complete the circle of their understanding of spatial theory and application. They have to conduct research, establish main points of attack, negatives and positives which may exploited in the argument, think on their feet and finally present their findings and conclusions. Debates incorporates several active learning tactics including group learning, peer teaching and increasing communication skills.

Issues and Problems Related to Technology Integration

The integration of this tool into a classroom environment presents multiple issues and potential barriers to overcome. The first major hurdle to cross for students is the concept of *map projections*. Map projections put a round earth on a flat map. Almost everyone is familiar with the concepts of latitude and longitude. Latitude is the measurement north and south from the equator. Longitude is the angular measurement east and west from the prime meridian (DeMers, 2000). Since the earth is generally depicted as an elliptical sphere, the students have to grasp the concepts involved in translating the earth's latitude and longitude into a flat map with x and y axes. The most straightforward way to do this is to simply replace latitude with x and longitude with y, keeping the same values. This will result in a flat map divided into degrees: 90 degrees south to 90 degrees north and 180 degrees west to 180 degrees east (Clarke, 2001). Consequently, "the map is now a map projection because the earth's coordinates (latitude and longitude) have been mapped or projected onto a flat surface" (Clarke, 2001, p. 56). This is only one of many map projections that cartographers have to choose from. Choice of a projection is based on what map properties the author of the map wants to preserve. Projections are usually grouped into four principal categories based on their preserved elements. The four projection categories are azimuthal, equal area or equivalent, equidistant, and conformal. Azimuthal projections retain accurate directional information, equal area projections retain accurate information about the size of an area, equidistant projections preserve distance information, and conformal projections preserve true shapes of lines (Chang, 2002). Other than conformal and equivalent projections, it is possible to have two preserved properties at one time (Chang, 2002).

Students are generally confused by projections because it is a complex concept that they must apply to their project data and to lab exercises, turning a 3-dimensional world into a 2-dimensional world. Uncertainty about which projection is most appropriate is usually overcome in several ways. In-class examples prove to be a very successful way of dealing with such confusion. Having students measure the distance from New York to Los Angeles using one projection from each major category shows them the significant differences among the results of inappropriate choices. For the final project, students must preserve some map information accurately to pose the spatial research question. The projection of the data must be known in order to identify what is being preserved. It is impossible to switch to a new projection without recognizing the old projection first. This is central to their projects. After they have created a spatial research question for their projects, they must give a written explanation of their chosen GIS analysis tools and projection. If this is completed early in the semester, feedback can be provided to them. Based on this feedback, there is less confusion and fewer mistakes as they make progress on their projects.

The second major hurdle typically encountered in Introduction to GIS is reliability of data. Many times the students will create their own data, find data on the Internet, or acquire data from an unknown author. Good *metadata* is intrinsically linked with every set of reliable spatial data. Metadata is essential data describing data. Metadata typically includes information such as author, map projection, accuracy assessment (if any), spatial resolution, creation date, update/change/edit date, information about the format, data reliability, and scale (Clarke, 2001). If students create their own data, many times they forget to record essential information about the data they produce. If students find data on the Internet, sometimes they will create a spatial research question around data that has either limited or no metadata; simply because it looks interesting when they view it as a map.

Students usually disregard warnings about data reliability until they are given unreliable data to work with as an exercise. To drive home the point, they are asked questions pertaining to data quality, scale, measurements they can and cannot make, and information preservation–things that cannot be answered unless they have access to a complete metadata file. This exercise, just in advance of their "project design" homework, has a very positive influence on their choice of data for their projects.

The third major barrier encountered with students is getting data into the computer. Numerous students want to create their own data based on facts in books or paper maps. Data from books usually requires lengthy data entry into excel spreadsheets. After hours of inputting data into spreadsheets, students are occasionally disappointed when they realize how little they have to analyze. Paper maps must be digitized. Digitization, depending on the complexity of the map, may take hours or even days to complete. It requires tracing every contour, line, point and polygon in the map. This is extremely painstaking work and if a person is careless, mistakes can be introduced into the map, reducing the reliability of the data. There are scanners available now that digitize parts of paper maps, but some features are missed, all must be labeled, and the final map must still be edited through much time and effort.

The solution to translating data from paper to digital map is to first advise students to seek data already in digital format. If that is not possible, a serious time management discussion is necessary to determine if the end result of their efforts is going to amount to acceptable project data. The process of finding suitable data or creating it is generally the most time-consuming step in the entire project process for most students. The key to avoiding student frustration is to clearly outline project goals, and give students a general semester timeline to follow with time dependent objectives to be achieved at regular intervals. A second way to avoid this problem is to provide students with a "data depot" that the professor has created for them to peruse and choose from. In my class I have a data depot as well as the data that is kept and managed by staff members of the GIS lab. The GIS lab has many types of data such as Dade county information, census data, and aerial photographs, for example. My own data depot contains data pertaining to coral reefs, African health studies, bathymetry, altimetry, and local roads. Many students, especially undergraduates, choose data from either of these resources.

The fourth and final major hurdle to overcome is difficult for both the professor and the students. It is definitively linking complex spatial theories and concepts with the GIS software. In most classes, facts, theories and concepts define the entire class. Here, not only are those elements being taught, but the student is forced to incorporate those elements into a digital environment and manage, manipulate, hypothesize and present results about those elements. It is extremely challenging. However, it is extremely rewarding to teach them to actively engage with the information and draw their own conclusions based on their own queries. With this repeated data display and manipulation, students realize what kind of cartographic representation best characterizes the purpose of their maps, queries and agenda. This therefore draws the link tighter between spatial theory and map creation.

Two Recommendations

Student comprehension of fundamental concepts revolves around repeated in-class hands-on exercises. Doing this compels students to actively learn and combine conceptual information with actual data analysis. The more spatial research queries they have to solve, the better they will become at incorporating the concepts into the software.

The second recommendation is not to overwhelm students with concepts too advanced for an introductory class. This is a temptation with motivated students. However, a comprehensive understanding of basic tools and theories will allow them to build upon a solid foundation of knowledge at a later date. That said, most motivated students request additional help in learning advanced techniques. Once taught, they usually teach others in the class on their own time. This is surprising but true.

Reflections Regarding GIS Integration

Enormous amounts of time and effort are required to teach GIS software in conjunction with spatial mapping history, theories and concepts. However, the graduate students who take the class are able to learn how to better analyze thesis data to earn their degree. Both graduate and undergraduate students, when finished, can put the course on their resume as a job skill. Currently, GIS job skills are in high demand. It is unusual that a single course can enhance a student's marketability. Therefore, even though this type of technology integration requires much time and energy on the part of the professor, it is well worth the effort.

Additionally, GIS teaches students that geography is a part of most disciplines, and that the concepts of space and place are important and interdependent. GIS can help show how social science, cultural studies and the physical environment are all linked. Hence, this technology allows for multidisciplinary interests. More and more, scientific investigations are not pigeonholed into physical or social categories, but are a combination of both. This trend encourages students to analyze the world as an entire system of human and non-human elements rather than as a sub-discipline of geography, biology or international relations. I believe this is a trend that will continue to flourish. For that reason, we should encourage our students to pursue skills that will better prepare them for analyzing spatial relationships in the world after college.

Footnotes

[1] ArcGIS, by ESRI, is the software used for this class at FIU but there are other GIS platforms: MapInfo, and TNTmips for example.

References

Arnheim, R. (1969). *Visual thinking*. Berkeley, California: University of California Press.

Ausburn, L., & Ausburn, F. (1978). Visual literacy: Background, theory and practice. *PLET*, *15*(4), 291-297.

Borzak, L. (ed.) (1981). *Field study. A source book for experiential learning*. Beverley Hills: Sage Publications.

Chang, K. (2002). *Introduction to Geographic Information Systems*. Boston: McGraw-Hill.

Clarke, C. (2001). *Getting started with Geographic Information Systems* (3rd ed.). New Jersey: Prentice Hall.

Creed, T. (1997). *Powerpoint, no! Cyberspace, yes*. National Teaching and Learning Forum, 6(4), 5-7.

Curtiss, D.C. (1987). *Introduction to visual literacy*. Englewood Cliffs, New Jersey: Prentice Hall

Debes, J. L. (1969). The loom of visual literacy: An overview. *Audiovisual Instruction*, *14*(8), 25-27.

DeMers, M. (2000). *Fundamentals of Geographic Information Systems* (2nd ed.). New York: John Wiley & Sons, Inc.

Edgerton, R., Hutchings, P., and Quinlan, K. (2002). *The teaching portfolio, capturing the scholarship in teaching*. Washington: AAHE Teaching Initiative–American Association for Higher Education.

Fransecky, R. B., & Debes, J. L. (1972). *Visual literacy: A way to learn, a way to teach*. Washington, D. C.: AECT Publications.

Haller, C.R, Gallagher, V.J., Weldon, T.L., and Felder, R.M. (2000). *Dynamics of peer education in cooperative learning workgroups*. J. Engr. Education, 89(3), 285-293 (2000).

Harmin, M. (1994). *Inspiring active learning: A handbook for teachers*. Virginia: Association for Supervision and Curriculum Development.

MacEachren, A.M. (1995). *How maps work–representation, visualization and design*. New York: The Guilford Press.

Mazzur, E. (1997). *Peer instruction, a user's manual*. Upper Saddle River, N.J.: Prentice Hall.

Meyers, Chet and Jones, Thomas B. (1993) *Promoting active learning: Strategies for the college classroom*. San Francisco: Jossey-Bass Publishers.

Silberman, M. (1996). Active learning: 101 strategies to teach any subject. Massachusetts: Simon and Schuster.

Smith, M.K. (2001). *David A. Kolb on experiential learning*. Encyclopedia of informal education. Retrieved April 12, 2004, from http://www.infed.org/b-explrn.htm

Theobald, D. (2003). *GIS concepts and ArcGIS methods*. Colorado: Conservation Planning Technologies.

Blended Learning and Geriatrics Education

Jorge G. Ruiz
University of Miami

Maria H. van Zuilen
University of Miami

Tara Kai
Florida International University
University of Miami

Michael J. Mintzer
University of Miami

Abstract

Implementing a competency-based medical school curriculum creates formidable challenges for geriatrics educators charged with the task of training physicians competent in the care of elderly persons. This chapter reviews the use of a blended-learning strategy for geriatrics education, which combines face-to-face methods with innovative e-learning technologies emphasizing self-instruction. Evidence indicates that the shift from a traditional approach centered on the teacher to one that is centered on the learner may result in improved satisfaction and cost-effectiveness as well as a more effective educational experience. The strategy proceeds sequentially, beginning with the identification of competencies and followed by the formulation of learning objectives, the development of assessment tools, and the ultimate construction of a blended-learning approach. At this stage one must first identify the appropriate instructional methods and then incorporate e-learning, which offers innovative and potentially cost-effective presentation and distribution methods. The effective delivery of a blended-learning strategy requires establishing a learning management system as the e-learning platform. Evaluation of educational impact follows a multilevel approach that gives attention to aspects of satisfaction, educational outcomes such as knowledge, skills, and attitudes, and cost-effectiveness.

Introduction

Medical schools face great challenges in preparing physicians who will provide competent care to growing numbers of elderly persons. Because medical schools and governing agencies have begun to mandate geriatrics training for undergraduate medical students, mainstreaming geriatrics into academic medicine has become paramount if educators want to ensure that all graduating medical students have the requisite knowledge, skills, and attitudes to care for an increasingly frail elder population. Both time and resources are scarce in education, and geriatrics medical school educators must seek out innovative cost- and time-effective educational approaches. This chapter presents a rationale and framework for one of these innovative educational approaches–a blended-learning approach to the geriatrics education of medical students that couples e-learning with traditional methods of learning to provide students and educators with flexible pathways for achieving desired educational outcomes.

This chapter first offers a rationale for using a blended-learning strategy in geriatrics education. A basic definition of blended learning and some examples of this strategy are provided. The role of learning management systems in delivering blended learning is examined, and finally, evaluation approaches are reviewed to assess the impact of the blended-learning approach.

Rationale for Using Blended Learning in Geriatrics Education

Nationwide, there is a shortage of health-care professionals trained to meet the special needs of older adults. While the number of elderly is increasing at a dramatic rate, it is estimated that that there will be 2,730 fewer geriatricians in 2004 than there were in 1998 and that this number will continue to decline in the coming decade (ADGAP, 2003a). In the face of this shortage of geriatrics expertise, medical schools and governing agencies have begun to mandate geriatrics training for undergraduate medical students. However, geriatrics educators and administrators face formidable obstacles to the development and implementation of geriatrics curricula that are student-centered and not teacher-centered, including the lack of curricular time, the scarcity of academic geriatricians, the increasingly decentralized nature of training venues, and the switch to competency-based education.

Efforts to enhance geriatrics training must first address the lack of curricular time devoted to geriatrics education. Medical school schedules are already overextended, with many disciplines vying for additional time, governing agencies setting more stringent educational requirements, and advances in the field of medicine that demand the incorporation of increasingly complex learning materials. There is an actual and projected deficit in the number of geriatric medicine faculty required for meeting the undergraduate educational needs in this field (Chiang, 1998). Unfortunately, this shortage will not likely be resolved soon despite the best efforts of academia and the federal government, although blending e-learning with traditional educational approaches may provide a pragmatic solution to the dilemma posed by limited resources. For example, e-learning components can be implemented before, and thus maximize the benefit of, a preceptor-led teaching activity. When students get such advance preparation, the limited time they then spend in the presence of a geriatrics faculty member can be concentrated on activities that foster humanistic and communication qualities, such as bedside teaching, and on preceptor-assisted activities, where faculty serve as role models for medical students in diverse training settings.

The expansion of training venues and spectrum beyond the acute care hospital to nontraditional settings such as nursing homes, hospice care, home-based care, and other community settings poses an additional challenge to designing and developing effective instructional strategies (ADGAP, 2003b). Considerable pressure has resulted in the frequent implementation of one-month-block mandatory rotations in geriatric medicine, but most of these experiences are inconsistent and discontinuous and require complex logistic arrangements (Warshaw, Bragg, & Shaull, 2002). Bringing e-learning into play will allow geriatrics educators to reach medical students across training venues regardless of distance and offers the further advantage of standardization of curricular components.

Another conundrum for geriatrics educators is the move away from the traditional curriculum toward competency-based education, which is by definition focused on the student, not centered on the teacher. Competency-based curricula seek to ensure that students attain specific knowledge, skills, and attitudes. The increased emphasis on accountability requires that geriatrics faculty devote substantial time and effort to restructuring their teaching and assessment strategies. But the shift to competency-based education can be aided by blended initiatives that offer students multiple ways to learn and excel, and possibly allow each student to tailor the learning experience according to an individual learning style.

From this discussion it is clear that a coherent strategic approach offers the best opportunity to meet the challenges now faced by geriatrics educators. Blended learning is a key ingredient to the success of such an approach.

The Concept of Blended learning

Background and Definitions

Blended learning is understood as the use of a traditional instructional strategy (usually face-to-face, teacher-led training) combined with "e-learning approaches" (Masie, 2002). As will be seen later, an important premise in this approach is that the instructional methods themselves (lectures, workshops, small-group discussions, problem-based learning, etc.) are independent of technology. The novel contribution of blended learning lies in its enhancement of instructional methods through the use of e-learning methods, first as presentation and then as distribution (Mantyla, 2001). For example, a teacher can elect to present a lecture (instructional method) either in person or by video (presentation methods). The students can then access the lecture in videotape format or through the Internet or an Intranet (distribution methods).

The basic concept of blended learning, or combining diverse teaching methods, is not novel in geriatrics education. Geriatrics educators have long used multiple delivery methods to facilitate learning by complementing classroom-based methodologies with assigned reading materials, group projects, or small-group discussions. However, in most cases, these strategies have been instructor-centered rather than learner-centered. A truly blended-learning approach to education potentially combines the best aspects of e-learning with the best of teacher-led training. Some examples relevant to geriatrics education include:

- A lecture about falls enhanced by a Web-based tutorial on gait and balance assessment

- A lecture on basic surgical techniques complemented by a computer-assisted simulation involving suturing
- A small-group discussion on environmental issues in the home augmented by an interactive CD ROM on home-safety assessments
- Bedside teaching with heart disease patients preceded by self-study with audiotapes of cardiac sounds
- A workshop about communication with elderly patients assisted by a videotape demonstrating how to deliver bad news
- A teleconference about diabetes mellitus in the elderly followed by a small-group discussion on treatment options
- A lecture about pressure ulcers incorporating a multimedia pictorial of various types of ulcers

In some cases, as shown by the above examples, the newly integrated e-learning component takes the form of self-instruction aided by technology. An important aspect of the blended-learning strategy is its reliance on newer presentation and distribution technologies that form part of what we know as the e-learning approach. One of the main advantages of e-learning is that it can be tailored to meet the needs of individual learners. A prime example is a Web-based tutorial on balance and gait assessment developed at the University of Miami School of Medicine that allows students to progress at their own pace, choose the parts of the tutorial they want to study, and revisit the content when desired (Ruiz et al., 2002a).

How Effective Are Blended-Learning Approaches?

Recent evidence suggests that blended-learning approaches that combine e-learning and traditional methods of instruction in a systematic way substantially increase productivity in corporate environments and seem to be a far more effective approach to training than single-delivery methods (Thomson Job Impact Study, 2003). Several studies have been conducted on the effectiveness of blended learning in the area of medical and geriatrics education (Andrews, Schwarz, & Helme, 1992; Elves, Ahmed, & Abrams, 1997; Hallgren et. al., 2002; Lipman et. al., 2001; McDonough & Marks, 2002; Mehta et al., 1998; Ruiz et al., 2002a; and Ruiz et. al., 2002b). Findings from these studies provide evidence of learner satisfaction, knowledge improvements, and cost and time- savings.

The future of blended learning depends largely on learners' reactions. Learner satisfaction with the components of blended learning is the first level of evaluation. If components are not well received, students will not be motivated to learn. While we have ample feedback from students on traditional learning methodologies, less research has been done on students' satisfaction with e-learning components. Mehta et al. (1998) conducted a prospective randomized study in which second-year medical students were assigned to either an experimental group exposed to both classroom and Web-based materials or a control group with access to the Web modules only after conclusion of the experiment. The post-test showed no significant performance difference between the two groups. While in the final survey only one percent of students believed that Web-based education could completely replace traditional teaching, 75 percent considered Web-based education an important additional educational resource. Ruiz and colleagues (Ruiz et al., 2002b) examined the usability of a computer-assisted training module on balance and gait assessment completed by second-year medical students. This module was used to prepare students for a preceptor-assisted, small-group session. Students favorably rated various aspects of the module's usability including navigation, pacing control, and ease of use of the self-tests.

The next step is to examine if a blended-learning strategy has additional benefits when compared with traditional methods or with just an e-learning approach. Several studies have documented that blended learning leads to greater improvements in knowledge than single-delivery methods. Fourth-year medical students receiving a "traditionally delivered" two-week geriatrics medicine curriculum were arbitrarily assigned to a group with or without a computer-assisted learning program. Students receiving the blended strategy demonstrated greater improvement in scores on a test covering the management of patients with dementia (Andrews, Schwarz, & Helme, 1992). Lipman et al. (2001) conducted a randomized study comparing a traditional classroom-based clinical ethics course with and without an Internet-based discussion component. Sophomore medical students' understanding of ethical analysis was significantly higher for those in the blended course than in the traditional course. In a randomized study by Stanford et al. (1994), first-year medical students who performed cardiac dissection and then viewed a computer-based application did significantly better when tested with computed tomographic and cardiac cadaver specimens than those students using the computer program or dissection alone. A Web-based interactive teaching tool complementing a packet of paper materials to assist first-year medical students' learning in a gross anatomy class was more effective in improving students' scores on anatomic landmark exams than paper materials alone (Hallgren et al., 2002). Elves, Ahmed and Abrams (1997) assigned third-year medical students to either a group with daily formal teaching or a blended-learning group that combined the face-to-face method with a computer-assisted training program in urology. When students were evaluated with a multiple-choice knowledge questionnaire, students in the blended-learning group performed significantly better than those in the face-to-face group.

Although incorporating e-learning in a blended-learning approach requires an initial investment of resources to develop the digital component, in the long run blended learning can save educators valuable time. McDonough and Marks (2002) conducted a randomized controlled study in which all participating third-year medical students had a 20-minute group lecture on basic concepts and historical aspects of exposure therapy for phobias. The students were then randomly assigned to either a computer-assisted training module or a face-to-face small-group tutorial, a traditional learning extension of the classroom-based approach. Students in the computer-instruction group scored as well on the multiple-choice-question post-test as those receiving the small-group tutorial. Although students reported more enjoyment with the small-group sessions, face-to-face teaching required five times more preceptor time (McDonough & Marks, 2002). The study by Ruiz et al., (2002a) found that students could learn some of the basics of balance and gait assessment by completing the computer-based tutorial. The tutorial freed up some of the geriatrics educators' time and allowed the time spent in small-group activities to be used more efficiently (Ruiz et al., 2002a).

New initiatives in blended learning are being conceived and implemented on more global levels. An example is the recently promoted international virtual medical school, a consortium of several medical institutions advocating the use of blended-learning approaches combining e-learning and face-to-face approaches to medical education (Harden & Hart, 2002). Such large-scale initiatives are particularly desirable in geriatrics education, because resource sharing among institutions can at least partially ease the dilemma of shortages of both geriatrics expertise and resources.

With blended learning, each element of the curriculum can be provided via the most suitable method. Given the variety of knowledge, skills, and attitudes required for competency in geriatrics, this flexibility in delivery is particularly important. Students can use computer-based training to learn about an area and then computer exercises and simulations to practice some elements of task performance. Time with instructors, other students, and patients will be used more efficiently, with the student asking questions, practicing task performance in the real world, and getting feedback and guidance from the instructor. Likewise, geriatrics educators can make better use of their time with students by focusing more on activities that foster role-modeling and humanistic qualities, essential instructional elements that are not well replicated, even by the most sophisticated computer software.

Blended Learning and Other Instructional Strategies

Until recently, most medical educators have used the traditional instructor-centered lecture approach as their predominant instructional strategy. Instructors are beginning to realize the limitations of the basic lecture format; it is usually a one-way process, in which the student is completely dependent on the instructor for the material presented and the timing of the presentation. However, traditional and blended learning are just two of the five major instructional strategies (see Table 12.1), which also include curriculum-integrated modular learning, distance learning, and electronic performance support systems. All five strategies have been used to a certain extent in the medical education field. Distance learning has not gained popularity, but electronic performance support systems (EPSS) have a part in clinical training in several academic institutions.

The instructional methodology itself–traditional methods, e-learning, or a combination–is not as vital as a well-developed instructional design process (Gagne & Medsker, 1996). Blended learning can be innovative and successful if it relies on systematic analysis of the learning task, the characteristics of the learner, and cost-effectiveness in relation to the best instructional strategy for specific learning objectives (Reiser & Gagne, 1983; and Gagne & Medsker, 1996). In contrast to the mainly teacher-centered traditional strategy, blended learning shifts the emphasis towards the learner. The e-learning component of blended learning accomplishes this goal by facilitating self-instruction. Empowering students with self-instruction approaches makes them active participants in the educational process and promotes lifelong learning (Spencer & Jordan, 1999).

Selecting the Right Components for e-Learning

There is no revolutionary change in the instructional/pedagogical methodologies of medical education when using the e-learning components of blended learning. The change lies in the incorporation of new e-learning presentation and distribution technologies. For example, a lecture (instructional method) can be offered face to face or via a teleconference (presentation method) and disseminated through CD/DVD-ROMs or the Internet as a Web cast (distribution method).

To achieve the right blend of e-learning and traditional methods, instructional designers must keep in mind the ultimate goal — student competence. A well-defined instructional design process (see Figure 12.1) is a prerequisite for reaching this goal. The competency-based approach to curriculum design begins with determining the competencies (knowledge, skills, and attitudes) a medical student is expected to acquire. The next step is to identify measurable learning objectives that would indicate whether a student has achieved those competencies.

After these objectives are established, a method of testing for those competencies is designed. Finally, educators design and develop a set of instructional activities so that students can achieve the identified objectives. It is precisely at this point that identifying the best-suited instructional methods becomes part of a rational blended-learning strategy. There are multiple pedagogical and instructional models that prescribe approaches to choosing the right instructional methods to achieve the specified learning objectives (Gagne, Briggs & Wager, 1992). After choosing a particular blended-learning strategy, a sequential step-by-step process may provide a structured approach to identifying the most coherent sequence of instructional, presentation, and distribution methods (see Figure 12.1), which in many cases will incorporate e-learning technologies.

Figure 12.1. Blended-Learning Flowchart.

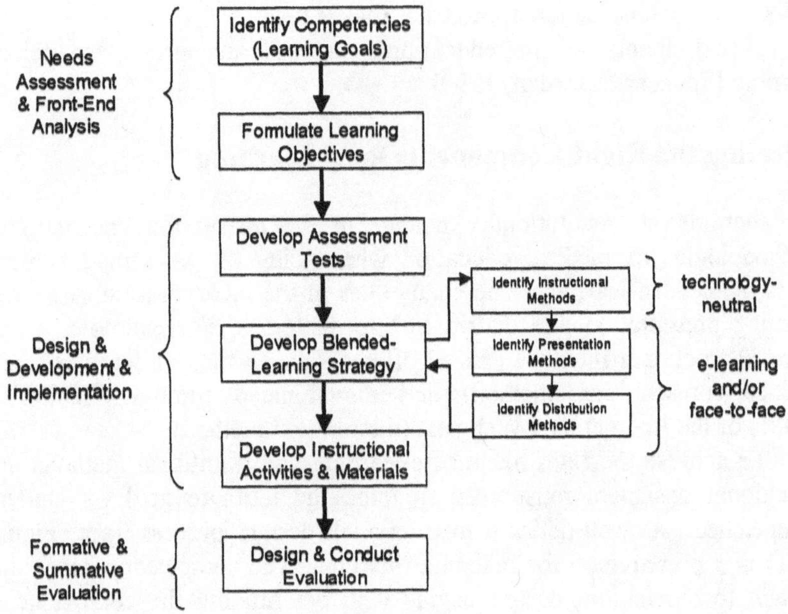

Table 12.1. Types of Instructional Strategies

e-Learning Approach		Description and Examples
Traditional Learning		Learning that is mediated by an instructor, typically in a classroom/clinical setting, and makes use of all the learning tools associated with the classroom environment. *A lecture to medical students about functional assessment in the elderly*
Blended Learning		Information-technology (e-learning) approaches combined with face-to-face training. *Geriatrics fellows are trained to perform cognitive screenings through the use of a face-to-face workshop and self-instruction with a learning object on the Mini-Mental State Examination (MMSE).*
Curriculum-Integrated Modular Learning		Part of a broad-based curriculum, with some modules completely managed through face-to-face delivery, and others delivered through information technologies. *During a medical student clerkship an ethics course is completely delivered face-to-face, whereas a legal medicine course is delivered through self-instruction with a computer-based training module. In contrast to blended learning, there is no attempt to weave traditional learning and e-learning in the same course.*
Distance Learning (Pure e-Learning)	Stand-Alone Synchronous Programs	Totally technology-based learning, but delivered at a prescribed time. *Physicians take an instructor-led online course on human resource management as part of an online master of business administration (MBA) program.*
	Stand-Alone Asynchronous Programs	Learner given complete freedom to choose to engage in a learning program, interacting only with technology. A *physician revisits an online module on the federally mandated Health Insurance Portability and Accountability Act (HIPAA) regulations as part of an online MBA (master of business administration) program.*
Electronic Performance Support Systems (EPSS)		Learning in the natural course of performing a job; learning incorporated in the workplace experience. *Internal medicine residents receive electronic alerts when prescribing exceedingly high doses of a medications to elderly patients.*

Instructional Methods

The term "instructional methods" refers to how the information is taught or what needs to be done to trigger learning (Piskurich & Sanders, 1998). Examples of instructional methods include lectures, small-group activities, question-and-answer sessions, problem-based learning, debates, case studies, tutorials, simulations, role-plays, active experimentation, exercises, and games. Instructional methods may lend themselves to a particular delivery method but are generally independent of technology. Role-modeling can be done live or virtually through the Internet. Choosing the right instructional method is a complex process that primarily depends on the intended measurable learning objective, which could be either knowledge, skill, or attitude (Gagne, Briggs, & Wager, 1992), but other factors must also be considered. Although any analysis of the considerable number of models and algorithms guiding this process is beyond the scope of this chapter, four essential factors may assist educators in the rational selection of the best methods (Gagne, Briggs, & Wager, 1992; and Stolovitch & Keeps, 2003):

Expected learning outcomes: The knowledge, skills, and attitudes we expect the student to achieve. Improving students' knowledge about dementia can be done through a lecture (face-to-face or Web-based). However, if the goal is to improve students' attitudes toward caring for demented patients, role-modeling of effective communication with demented patients and role-playing may be more effective.

Feasibility: This consideration refers to whether a desired method is viable given existing resources and limitations. The preferred instructional method may be a small-group discussion, but some institutions may lack the trained faculty to facilitate small-group sessions.

Acceptability: This concept refers to whether the environment and culture of the academic institution and the characteristics of the learners are receptive to this instructional method. Although more institutions are implementing problem-based learning activities, many institutions remain reluctant to accept this instructional method.

Economics: This factor refers to whether the methods are the most cost-effective and cost-efficient given institutional resources. Providing Web-based instruction to medical students may not be a viable proposition if the institution does not provide enough computer and Internet access.

Presentation/Delivery Methods

How information is presented to learners (i.e., the instructional method) is termed the presentation, or delivery, method (Piskurich & Sanders, 1998). Some examples include face-to-face (instructor-led), computer-assisted training (CAT), electronic performance support systems (EPSS), interactive TV, chat, audio, teleconferencing, virtual reality, video, multimedia, online help, electronic bulletin board, and reusable learning objects.

Distribution/Media Methods

The medium used to deliver information to learners is the distribution, or media method (Piskurich & Sanders, 1998). Some examples are live (classroom-based), cable TV, closed-circuit TV, audiotape, videotape, CD-ROM, DVD-ROM, Internet, Extranet, Intranet, Local Area Network (LAN), Wide Area Network (WAN), Wireless Fidelity Network (WiFi), satellite TV, tactile-gear simulator, telephone, voicemail, World Wide Web (WWW), floppy disk, e-mail, newsgroup, and listserv. The selection of the presentation and distribution methods is guided by the same factors outlined for deciding on instructional methods. In order for blended learning to succeed, new e-learning presentation and distribution methods must be readily accessible to learners and faculty.

A Learning Management System for Geriatrics

Learning Management Systems (LMS) are critical ingredients for the success of blended-learning initiatives. A LMS is software (running on an Intranet or the Internet) that supports an educational institution in the delivery, management, and administration of learning across the institution. In the context of blended learning, an LMS facilitates the delivery of e-learning content to students and faculty. However, an LMS can serve functions other than the delivery of online instruction. It can simplify and automate administrative and supervisory tasks, track the impact of education on students' achievement of competencies, and act as a repository for instructional resources that can be available to learners and faculty twenty-four hours a day. Some of the key features of an LMS are presented in Table 12.2. The number of commercial LMS systems is increasing rapidly, with between 200 and 300 software packages currently available. Well-known examples include WebCT, Blackboard, Top Class, Angel and Desire2learn. The LMS used for delivery of our blended-learning strategy provides a platform to deliver geriatrics e-learning materials to medical students, geriatric medicine fellows, and nurses in long-term care (Ruiz et al., 2002b).

Evaluating Blended Learning

Evaluation is an important part of any educational program (Wilkes & Bligh, 1999), but programs that involve new information technologies and processes require more evaluation and justification to gain acceptance than traditional instructor-led methods (Friedman & Wyatt, 1997). However, blended learning is an educational intervention, and its evaluation can be confounded by multiple concurrent factors (Wilkes & Bligh, 1999). A multilevel approach to evaluation should consider three important areas: satisfaction, learning, and cost.

Satisfaction: For blended learning to be effective, its different components must be well received by the learner and other stakeholders. Satisfaction depends on usability, efficiency, and effectiveness. The success of any educational resource begins first with how comfortable faculty and learners feel when they use it. Evaluating usability, or user friendliness, may or may not be important with traditional methods of learning, but the task is imperative for new blended-learning methods that often incorporate unproven information technologies. For example, if students are asked to complete a Web-based training module, the evaluation should consider questions about accessibility, ease of navigation, and control of pacing of the module. Ratings of usability are also influenced by the efficiency of different components of the blended-learning process. How much effort (usually measured in time) is required by students and faculty to complete a component? Time-consuming face-to-face teaching sessions might tax faculty endurance, but on the other hand, overly long Web-based modules might result in student disengagement. Finally, the perceived effectiveness of blended-learning components in achieving the specified learning objectives further contributes to satisfaction. An educational component that is not usable, effective, or efficient cannot possibly be evaluated for higher educational outcomes, such as learning.

Table 12.2. Features of Learning Management Systems

Monitoring and tracking of student performance
Testing and reporting results
Management of student records
Evaluation for competency improvement (survey and test builders)
Course scheduling and organization
Online student enrollment and administration
Knowledge, skill, and attitude assessment
Course delivery (asynchronous and synchronous)
Organized access to digital libraries (including learning objects)
Integration of third-party content including multimedia and authoring tools
Collaboration features (e-mail, bulletin boards)
Database support
Personal tools
Support standards (SCORM, IMS, AICC)

Learning: The focus of blended learning in the context of a competency-based geriatrics curriculum is on the attainment of knowledge, skills, and attitudes in the care of elderly persons. It is critical to evaluate the utility of blended-learning approaches to determine if actual learning has occurred. After competencies and specific learning objectives have been identified and appropriate assessment tools developed, we can measure the impact of blended learning on the knowledge, skills, and attitudes of learners. Medical schools have traditionally relied on multiple-choice exams and preceptor ratings for student assessment. Although these methods have become familiar and comfortable for students and faculty, there are major weaknesses in their use for evaluation of learning. Multiple-choice exams have limited professional authenticity. Test items cannot adequately assess the complexities of taking care of a frail elderly person. Preceptor ratings generally have low inter-rater reliability (Newble & Cannon, 2001).

Table 12.3. Assessment Methodologies

Assessment Methodologies
Standardized written examinations (multiple-choice, short-answer, essay)
Oral examinations
Paper-and-pencil questionnaires
Faculty/preceptor ratings
Objective structured clinical examinations (OSCEs) of patients
Real-time or video performance observations
Computer simulations and self-assessments
Case presentations
Patient logs
Structured interviews
Educational portfolios
Medical record audits
Ethnographic observations

A blended strategy can be extended to student assessment to counter some of the weaknesses of individual approaches. Table 12.3 lists a variety of assessment strategies available to educators, which can be tailored to the specific desired educational outcomes. Using more than one method of evaluation is appropriate in many cases. For example, students can perform a self-assessment on the scoring of a mental status exam of an elderly confused patient via a Web-based module. This exercise allows for a degree of standardization, assures that all students have a similar patient exposure, and limits (or eliminates) the time a preceptor needs to be present. At a later time, the exercise can be reinforced when a preceptor assesses students' performance of a mental status exam on a real or standardized patient.

Cost: Comparing the monetary benefits of a blended-learning approach, to the costs of the traditional learning program is another important component of evaluation. As mentioned earlier, a potential benefit of blended learning is its ability to reduce costs by redirecting limited resources and improving efficiency for educators and administrators. Economic outcomes that could potentially be measured are the number of man-hours originally dedicated to preceptor-assisted activities that are now reassigned to e-learning components, the expansion of the amount of content provided without additional faculty effort, and the reduction in administrative costs for monitoring and tracking student performance and competencies accomplished by switching to the automated functions of a LMS.

Conclusions

In this chapter, the authors have proposed that the use of a blended-learning strategy combining face-to-face methods with innovative e-learning technologies might assist geriatrics educators in implementing a competency-based medical school curriculum. A blended strategy shifts the emphasis from the traditional teacher-centered approach to one that is centered on the learner and can potentially enhance the quality of the educational experience and improve student satisfaction and cost-effectiveness.

A rational instructional approach begins with the identification of competencies, followed by the formulation of learning objectives, the development of assessment tools, and the ultimate construction of a blended-learning strategy. This strategy requires first the identification of appropriate instructional methods and then the incorporation of e-learning, with innovative and potentially cost-effective presentation and distribution methods. A learning management system as the e-learning platform allows the delivery of this strategy. A multilevel evaluation will include among other things, evaluation of satisfaction, knowledge, skills, and attitudes, and cost-effectiveness.

References

Andrews, P. V., Schwarz, J., & Helme, R. D. (1992). Students can learn medicine with computers. Evaluation of an interactive computer learning package in geriatric medicine. *The Medical Journal of Australia, 157*(10), 693-695.

Association of Directors of Geriatric Academic Programs (ADGAP). (2003a). *Longitudinal study of training and practice in geriatric medicine training & practice update: Geriatricians and geriatric psychiatrists (Report)*. New York: Association of Directors of Geriatric Academic Programs (ADGAP).

Association of Directors of Geriatric Academic Programs (ADGAP). (2003b). *Longitudinal study of training and practice in geriatric medicine training & practice update: Medical student training in geriatrics at the beginning of the 21st Century (Report)*. New York: Association of Directors of Geriatric Academic Programs (ADGAP).

Chiang, L. (1998). The geriatrics imperative: meeting the need for physicians trained in geriatric medicine. *Journal of the American Medical Association, 279*(13), 1036-1037.

Elves, A. W., Ahmed, M., & Abrams, P. (1997). Computer-assisted learning; experience at the Bristol Urological Institute in the teaching of urology. *British Journal of Urology, 80*(Supplement 3), 59-62.

Friedman, C., & Wyatt, J. (1997). *Evaluation methods in medical informatics*. New York: Springer Verlag.

Gagne, R., Briggs, L., & Wager, W. (1992). *Principles of instructional design* (4th ed.). Belmont: Wadsworth/Thomson Learning.

Gagne, R., & Medsker, K. (1996). *The conditions of learning: Training applications*. Belmont: American Society for Training and Development.

Hallgren, R. C., Parkhurst, P. E., Monson, C. L., & Crewe, N. M. (2002). An interactive, Web-based tool for learning anatomic landmarks. *Academic Medicine, 77*(3), 263-265.

Harden, R. M., & Hart, I. R. (2002). An international virtual medical school (IVIMEDS): the future for medical education? *Medical Teacher, 24*(3), 261-267.

Lipman, A. J., Sade, R. M., Glotzbach, A. L., Lancaster, C. J., & Marshall, M. F. (2001). The incremental value of internet-based instruction as an adjunct to classroom instruction: a prospective randomized study. *Academic Medicine, 76*(10), 1060-1064.

Mantyla, K. (2001). *Blending e-learning*. Alexandria: ASTD.

Masie, E. (2002). Blended learning: the magic is in the mix. In A. Rossett (Ed.), The ASTD E-Learning Handbook (pp. 58-63). New York: McGraw-Hill.

McDonough, M., & Marks, I. M. (2002). Teaching medical students exposure therapy for phobia/panic — randomized, controlled comparison of face-to-face tutorial in small groups vs. solo computer instruction. *Medical Education, 36*(5), 412-417.

Mehta, M. P., Sinha, P., Kanwar, K., Inman, A., Albanese, M., & Fahl, W. (1998). Evaluation of Internet-based oncologic teaching for medical students. *Journal of Cancer Education, 13*(4), 197-202.

Newble, D., & Cannon, R. (2001). *A handbook for medical teachers* (4th ed.). Dordrecht: Kluwer Academic Publishers.

Piskurich, G., & Sanders, E. (1998). *ASTD models for learning technologies: Roles, competencies, and outputs*. Alexandria: ASTD.

Reiser, R., & Gagne, R. (1983). *Selecting media for instruction*. Englewood Cliffs: Educational Technology Publications.

Ruiz, J., Granville, L., Ribera, V., Kai, T., Smith, M., van Zuilen, R., & Roos, BA. (2002a). Computer-assisted training in gait and balance assessment. *Journal of the American Geriatrics Society*, *50*, S73.

Ruiz, J., Kai, T., Smith, M., Granville, L., Williams, C., Mintzer, M., & Roos, BA. (2002b). *GeriU*, the Online Geriatrics University (www.geriu.org): Enhancing geriatrics education through web-based learning. *Journal of the American Geriatrics Society*, *50*, S72.

Spencer, J. A., & Jordan, R. K. (1999). Learner centred approaches in medical education. *British Medical Journal*, *318*(7193), 1280-1283.

Stanford, W., Erkonen, W. E., Cassell, M. D., Moran, B. D., Easley, G., Carris, R. L., & Albanese, M. A. (1994). Evaluation of a computer-based program for teaching cardiac anatomy. *Investigative Radiology*, *29*(2), 248-252.

Stolovitch, H., & Keeps, E. (2003). *Engineering effective learning toolkit*. San Francisco: John Wiley & Sons, Inc.

Thomson Job Impact Study, Final Results. (2003). *The next generation of corporate learning: Achieving the right blend*. Naperville: Thomson NETg, Inc.

Warshaw, G., Bragg, E., & Shaull, R. (2002). *Geriatric medicine training and practice in the United States at the beginning of the 21st Century (Report)*. Las Vegas: Association of Directors of Geriatric Academic Programs.

Wilkes, M., & Bligh, J. (1999). Evaluating educational interventions. *British Medical Journal*, *318*(7193), 1269-1272.

Technology as Enhancement Tool for Professionalization and Cognition: The Effects of Ultra-Portability and Meta-Cognition Strategies

Mark D. Szuchman
Florida International University

Abstract

A graduate seminar in the field of history incorporated the use of technology to enhance student performance specifically in the course and more broadly in their graduate preparation. The first goal was to acquire data in mobile fashion that would be easily accessible and subject to manipulation and updates. The second goal was to use technology to facilitate meta-cognition processes and skills required by the profession. Students were required to take notes on assigned readings using Palm Pilots and keyboards. They subsequently uploaded their notes to Nota Bene, a desktop application designed to handle bibliographic management, data mining, and academic writing. In-class discussions focused on conceptualization skills, data mining and bibliographic accumulation. The final paper, a literature review, accomplished two goals: it brought together technical skills with substantive content, and served as a means to demonstrate the students' mastery of bibliographic resources and authors' theoretical contributions.

Introduction

This work is designed to share the motivation, methodology, and mechanisms involved in re-engineering advanced courses in history by taking advantage of new technologies and adapting them to goals involving both content and cognitive processes. I make use of the plural, that is, "technologies," not as a discursive and self-referential affectation, but as an indication of an approach to teaching and learning informed by an awareness that pedagogical objectives must drive our discrimination from among arrays of technological solutions. I should make it clear from the outset that the use of computer technologies in the context of this work does not apply to content; that is, often, educational technologies are invoked in courses in order to provide students with text documents or audio-visual aids available on the Web. Web-based content resources in the field of History are highly variable in quality and availability, with the North American and European content holding pride of place.[1] Beyond content, computer technologies facilitate course management (e.g., grade rolls), handouts to students (e.g., graphs or tabular data), or presentations on a screen projector (e.g., the principal lecture points, lecture outlines, etc.). My objectives in re-designing my seminar were aimed not at the students' ability to receive information, but, instead, for students to learn new computer skills that would reinforce their abilities to process information. This involved a paradigmatic transition aimed at sharpening the distinction between students as passive data observers in contrast to active processors of data.

The approach I have taken to the design of my courses is not novel. Indeed, it is based on the by-now traditional notion that students learn best, first, by understanding the learning process, second, by adopting the disciplines' conventions, and, third, by linking community-building among members of the class with the learning process. The technological tools that I selected retain the same pedagogical constructs that have concerned me for some time–indeed, well before the entry of technologies into the teaching equation–but such constructs, in turn, have been shaped and enhanced by the potentialities of selected technologies. Finally, I would stress that my course's field of study–Latin American History–does not present any limiting idiosyncrasies. The course design and technological adaptations discussed here can respond equally well to a broad range of disciplines cutting across the humanities and social sciences. In this respect, content is secondary in this discussion.

The Paradox of Teaching and Learning

I will now present the pedagogical issues that have concerned me over the years, particularly in the context of advanced levels, which, in history, normally take place in seminar settings. A basic paradox between what we do as professors and what we expect from our students came into sharper focus when several of our better graduate students failed one or two of the written sections of their doctoral comprehensive examinations in Latin American History. This occurred over the course of a couple of years. I was puzzled about how it came to be that individuals who had been considered good students, and who had performed well in their courses, who had written good papers and who had provided strong contributions during seminar discussions–in sum, who had fulfilled all the criteria for success in a graduate history program–came to such an abrupt halt in progressing toward their degrees.

Thus, my concerns originally focused on the shortcomings in student performance in the specific context of performance in the comprehensive examinations leading to doctoral candidacy. As I began to read in the field of student learning–and I would not claim great expertise in this field–I became increasingly convinced of a curious disconnect between our teaching, on the one hand, and the performance expectations we have of our students, on the other. I will further discuss recent technological developments that sparked my imagination, helping to shape a vision of student participation in the learning process. I will then describe the construction of the course. I use the term "construction" because, with hindsight, it became a process of putting together components designed to fit together, framing and supporting a structure of learning.

The Seminar Setting–Hidden Traps

The seminar is the common setting for advanced courses in history; rarely are graduate courses offered in lecture format. The seminar presents a significantly more challenging setting for both instructor and student. Further, it is an environment filled with contradictions, a significant amount of implicit expectations, and a great deal of apprehension. The contradictions present themselves at the first seminar meeting. To start, students are typically told that a significant component of the grade–often at least fifty percent–will be determined by the quality and nature of their participation in the discussions that typically define the substance of seminars. But what type of participation is possible for students who often are taking a subject area to which they have had little or no exposure. It is not unusual for the more advanced graduate students to comment on their first term taking courses in graduate school in many fields of the humanities and social sciences and noting that the experience represented a sort of intellectual baptism by fire. It is not uncommon for the instructor to distribute a list of introductory texts to the field and inform the students that those without sufficient general knowledge should immediately read one or two of them in order to equip themselves with some context in the more discrete topics covered by the seminar. A Malthusian world awaits many students starting out in graduate school. Does this have to be? Perhaps not, but it is a reality that has served many generations of historians (and other social scientists and humanists) well, and is likely to continue. After all, it provided an effective mechanism of teaching and learning for Aristotle and his students.

The seminar setting is designed to develop students' intellectual autonomy, not only in History, but also in a wide range of professional endeavors. The seminar's goals focus on the cognitive and social realms and include the:

- development of imaginative and creative thinking
- expansion of a critical and informed mind
- development of rigor and understanding of causality
- socialization unto the norms, practices and conventions of the field of study to which the course belongs (Preston, 1996).

These goals indicate that we have become increasingly aware that the process of learning is at least as important as the content we learn, and that content is, in turn, conditioned by the cognitive skills we bring to bear on the process of understanding. Indeed, thousands of hours have been dedicated to the challenge of understanding, conceptualizing and relating data to judge by the over 95,000 web sites found by an internet search of "note-taking." Even after discounting for commercial products, one is left with an extraordinary number of instances of people trying to make an impact on the way students–indeed, any researcher–improve their summary and conceptualization of materials read and heard. Here we find yet another contradiction, in that it is not clear that, normally, professors of history and cognate fields, such as sociology, political science, etc., have taken pains to become actively engaged in augmenting the development of such skills. This is all the more remarkable insofar as there is a growing recognition that the reasoning skills attendant history courses carry implications for areas of endeavors far afield, affecting such diverse areas as policy-making, nation-building, community development, inter-generational communications, and belief-systems (Winburg, 2001, p. 51). To be sure, matters are slowly changing. Only most recently has the history professoriate become involved in thinking about the historical learning process, perhaps motivated by a perceived threat to the profession from the more applied fields of knowledge, perhaps by an awareness that the historical mode of analysis requires early cross-disciplinarity.

As Leon Fink, a vice-president of the American Historical Association, noted recently about the efforts of the Association's Teaching Division to reach out beyond the ivory tower,

> the common intellectual puzzles we confront in approaching our students [are,] first, how do we define for ourselves the essential skills or ways of thinking *basic* to the historian's craft? And second, how do we best package such gifts for our students? Such questions are no more the province of the academy than the school, and they ought to be part of the core curriculum in both social studies teacher education programs as well as history doctoral training (Fink, 2001, p. 235).

Fink and others in the historical profession, catalyzed by the efforts of Diane Ravitch and her colleagues in education and cognitive sciences who have questioned the modes of teaching and learning history, understand better the entwined nature of how students learn and how we should teach (Fink, 2001, p. 236). Yet, on the whole, one finds in the professoriate a much greater willingness to change the content of course coverage–new themes, additional group representations, "culture wars" engagements–than to reflect on our contributions to shape the cognitive interactions. I agree with Sam Wineburg's observation that "the traditional view, in which knowledge goes from the page of the text to the head of the reader, is inadequate. But the metacognitive view, in which knowledge is constructed by students questioning themselves about a fixed and friendly text, is equally inadequate" (Wineburg, 2001, p. 83).

An additional contradiction presented by the seminar setting involves the need for instructors to be aware that they are called upon to intervene judiciously, and to resist the temptation to take over the discussion. Whether to fill a void in the students' participation, or to re-direct the flow of discussions toward an unstated goal, instructors often take over in ways that may lead to additional instances of what might be called "instructor hegemony." In turn, this leads to the continued student passivity that is often a source of professors' complaints. But, how does an instructor fulfill the obligation to act as the intellectual leader and shape the direction of discussions so as to yield meaningful understandings, yet without hegemonizing? This is still another underlying dilemma of the seminar setting.

Origins of Course Objectives

I first re-designed a graduate seminar during the Fall of 2000 in preparation for the offering in the Spring of 2001–which dealt with family history in Latin America. Subsequently, I made additional design changes in a seminar I co-taught the subsequent Spring with a colleague whose specialty is in European family history. Not only was the second version of the seminar designed to capitalize on the comparative framework to the relationships between family and state in Europe and Latin America, but it also permitted a colleague to observe and comment on the technological mediations.

The re-design was the consequence of thinking about the problem of doctoral students who had presented themselves to the comprehensive candidacy examinations (with a concentration in Latin American History) but who failed at least one of the questions. This occurred to several students over the course of three years. All the students had demonstrated strong academic records. I could vouch for two of them, who had done very good work in my seminars. What had gone wrong?

The Invisible Paradox: We Teach, They Wander

In an era of increased specialization, faculty members in social science and humanities have too frequently narrowed the scope of seminars. However, the monographic quality of seminars runs counter to the objectives inherent to comprehensive examinations, which involve conceptual breadth and broad comparability. The very term "comprehensive" indicates the expectation for inclusionary modes of inquiry and response. Indeed, we penalize students papers most heavily for lacking in analytical skills, for insufficient conceptualization, or for excessive fact-telling that is not mitigated by careful analysis–a term that is hardly self-evident, insofar as it is contextual, vastly under-explained by the professor, and little understood by the student. One of my objectives, thus, included closing the gap–or better still–creating the nexus between the relatively narrow scope of a monographic seminar and the broader challenges presented down the road by a comprehensive examination.

Closing the Gap

I became convinced that, however well we teach content in our monographic seminars, we do not normally dedicate much effort in the direction of building mechanisms capable of aiding our students to accumulate the building blocks embedded in individual courses with the goal of constructing stronger theoretical and comparative arguments as part of their overall development. That is, we do not normally include in our class sessions the accentuation of conceptual skills, analytical frameworks or comparative avenues. When we do so, it is more often by way of understatement than emphasis. As faculty, we need mechanisms to help us to not lose sight of the concept-building objectives and that involve both the means and the oversight of students' annotations regarding their readings and class discussions. We *assume* that what they read and discuss undergoes a cogent and useful process in their note-taking. In practice, however, we cannot know if this is the case because we do not normally build into the course the mechanisms to account for student annotations. Score this, too, as another puzzling contradiction, or gap, between what we teach and how we practice research. One of the basic mechanisms in conducting historical research (or research in cognate fields), involves annotating our readings, taking notes that help us to process the data that we accumulate. And yet, we normally do not teach with these skills in mind. How should we, as members of the faculty, build skills that are normally rewarded in our field of study?

Categories, Concepts, Comparisons, Capaciousness

The field of History calls for abilities to create, classify, order and understand categories. We share this fundamental objective with colleagues in cognate disciplines ranging from sociology to philosophy, from anthropology to religious studies, from literature to psychology. Our friends in the physical and life sciences, too, share an emphasis on categories–indeed, they bring us to an understanding of hierarchy in life forms, molecular configurations and other classifications–but I put these scientists aside, recognizing that their forms of thinking and data processing require a different optic, a different mode of inquiry, and significantly different modes of observation.

Categories are essential to understanding phenomena. A clear awareness of categorization as a tool for ordering empirical data serves to enhance research and modes of inquiry–that is, the associated cognitive processes are essential. Categories lead to hierarchy and forms of relationships, and no set of relationships is more complex than those created by humans, the fundamental unit of analysis for historians. More challenging for students still is the goal of discerning relationships from among people whose behaviors are not observable to us directly, but, rather, mediated by their written records or the written words of others. Historians are among those analysts whose understanding of worlds gone past can make sense only by the creation of categories that inform certain types of relationships in order to facilitate analysis and explanation. As researchers and students, our objectives lie within the realm of observing fragments of human interactions in the realms of politics, culture, along with social and economic dimensions so that, during the trek (the particular course or the research project) we may patch the fragments together to form patterns serving us to recognize the whole. We work, in other words, in granular fashion, from the bottom up.

But just as important a goal, historians also work with theory. We are motivated by initial hypotheses about the whole, which, in turn, is followed by the sleuthing process in search of the relevant components in a process of observation and formulation. Normally, the bottom-up (data gathering) and the top-down (data processing) modes of constructing the project operate in symbiosis (Palmer, 1975). We should note that the simultaneity of top-down and bottom-up modes of inquiry and interpretation is hardly peculiar to historians (Palmer, 1975). In such academic disciplines, categories and hierarchies are obviously based on words, and of particular importance in the teaching-learning context, the appropriateness of words as tools that encapsulate meaning. To know the various meanings of words, and in particular, words that resonate with a rich array of historical behaviors, that is, that involve concepts capable of dissecting complex human interactions, represents a teaching challenge: we need to emphasize skills that give students the opportunity to understand what they read in order to make class discussions more effective.[2]

Because historians work normally within discrete periods and regions, they plot their categorization of evidence along spatial and temporal axes. Historians manage large amounts of data, which can undergo both quantitative and qualitative analyses. But quantitative analysis, by itself, is not the field's objective, and even the most committed quantitative historian will tell us that quantitative data are important tools to strengthen reasoned and qualitative arguments. Thus, qualitative analysis and argumentation of historical conditions represent our ultimate objectives. In order to do this well, we are required to read a great deal and meet the challenges of imposing lines of order and relationships, data categorization, and access to data. For every additional manuscript in the archive, article, book, book review, lecture, or web-based material we read, we are both enriched and placed deeper into an increasing swarm of data and ideas. For historians, along with other social scientists and humanists, the challenge has long been–even before the so-called "information age"–to control and retrieve text-based data. Our task is fundamentally cognitive, but it is also highly dependent on spatial conditions, as an information scientist asked, "how does the historian manage–through spatial placement and physical and mental labeling–the large quantities of text that are gathered" (Case, 1991, p. 660) in the creative process?

One of the most useful and traditional tools used by historians to store data and ideas is the note card. As indicated earlier, good note-taking skills present us with an interesting paradox: one of the essential skills used to conduct our business is not formally addressed in the higher education classroom and it is barely addressed in the secondary school setting. Computer technologies have held the promise of facilitating note taking. We see the realization of this promise in the form of students in the library typing into their laptops the materials needed for their courses and academic projects. But, in fact, they are not usually advancing the goals of ordering, categorizing, or conceptualizing. To be sure, they are operating in a digital medium, but in the end, they are doing little more than typing. Furthermore, the very ease of entering data, in this case, words into the computer, accelerates the process of accumulation and increases the challenge of locating the material. It often takes less time to find information on the Web than in one's own computer. Thus, computer technologies present us with yet another paradox: the more data we accumulate on our computers, the more inefficient we become in retaining mastery over the materials, complicating the goals of categorizing, conceptualizing and comparing. As wordsmiths and qualitative data analysts, we have become owners of more capacious storage facilities but not of greater cognitive capacities.

To sum up, I embarked on re-designing my seminar on the basis of enhancing students' skills by facilitating conceptualization, by employing the seminar setting as a community-building venue, and by using technologies to achieve my objectives. I would focus on making explicit the meta-cognitive dimensions of my agenda, require computer skills and provide the technological means to arrive at a lasting change in the students' approach to reading and discussions. Finally, such changes would carry beyond the course into others, whether in history or cognate fields in the social sciences and humanities (Steiner, 1996, p. 116–117).

Technological Mediation

I have long required students in my seminars to bring to class two copies of their notes on the weekly assigned readings. At the end of each class meeting, I would collect one copy and leave the other for students to select randomly and keep. This device virtually eliminated non-compliance with reading assignments and allowed me to have an "early warning system" by noticing problems with how the readings were being understood. In addition, my system offered students opportunities to observe how their peers treated the same materials. For some, this became a validation of their note-taking skills and understanding of the assigned reading, while others had opportunities to improve theirs. My re-design did not alter fundamentally my teaching process or seminar management or procedures. Instead, I capitalized on my existing modus operandi, which provided the starting point, and enhanced it through technological mediation.

New Developments In Hardware

Over the course of the 1990s, my students would increasingly turn in copies of many of their notes as word-processed documents. Occasionally, if they did not have the time to transcribe their notes when they read materials in the library (which is the norm), I received them in handwritten manner. However, the development of the portable and foldable keyboard for the Palm Pilot in 2000 provided me with the basis for a sea change in data entry practices, as the device expanded significantly the potential for ultra-portable computing, particularly in the area of text entry.[3] Using the Palm Pilot with the Palm keyboard, it was not long before

I stopped taking a laptop to the library for my reading, research and note-taking. The time was fast approaching when all my students would be entering their notes in digital fashion.

Software

Within the span of a few months after the Palm keyboard was introduced, freeware and shareware applications for word-processing with the Palm Pilot appeared on the market. The availability of WordSmith, published as shareware by the Blue Nomad Corporation, signified the arrival of a sophisticated word-processing application at a very reasonable cost. Files were now capable of being saved in Rich Text Format, facilitating virtually universal file exchange among Palm Pilot users, on the one hand, and between Palm Pilots and computers, on the other, while retaining document format and print modes. As a result, my students would no longer need to bring their laptops to the library to enter data and produce their notes. In fact, with appropriate hardware, they can even produce documents directly from the Palm Pilot using printers with infrared capabilities, bypassing the need to use a computer for printing purposes.

Beyond the technological developments that made possible ultra-portable data entry, use of the Palm Pilot, in turn, opened the door to a cognitively powerful learning process. This was achieved by joining the Palm-centric document to Nota Bene, a specialized academic software suite designed for the desktop, which focuses on the academic activities of researchers especially in the social sciences and humanities. Thus, WordSmith documents, created in the Palm Pilot, are brought to Nota Bene during the synchronizing process, for subsequent use in powerful ways.

But first I had to meet my greatest challenge, which consisted of a population that almost invariably used Word, the Microsoft product, for word processing. Changing word processing applications is one of the more difficult challenges to overcome. And yet, many of my objectives in enhancing students' skills and habits of mind required the use of new software. How then, to effect change? Fortunately, I was able to call upon my university to provide a grant to fund the hardware, which the students would return at the end of the term, and the Nota Bene suite, which students who succeeded in the course would get to keep. The students would be responsible for acquiring WordSmith for the Palm Pilot, a modest cost. It now became a matter not simply of requiring use of the application, but also of presenting students with a realistic sense of the advantages of using the application in order to meet not only the course's objectives but also the objectives of the students in the course of their academic and professional development. In the next section, I will discuss the methods by which I went about moving the students to embrace the use of this specialized software.

Nota Bene is an integrated suite, operating in the Windows environment, that combines four applications. It includes word processing, bibliographic and reference database management (Ibidem), free-form hypertext searching and retrieving (Orbis), and a general-purpose, non-bibliographic database program (Ibid Plus) which I will not discuss insofar as it is not used in the course. All applications in the Nota Bene suite share a common interface, facilitating user training. The intricacies of layout that form part of academic styles are largely automated: the suite comes equipped with built-in styles for the MLA, APA, Chicago, and Terabian style manuals, implementing the specific layout rules of major academic disciplines and journals. By requiring papers to be produced with the appropriate academic style (in our case, it is normally the Chicago Manual of Style), the software's ability to automate the process serves as a powerful socializer and an important tool for enhancing the sense of professional standards among advanced students. This is a dimension of student preparation that is too often ignored in the educational process. I refer to the conventions of writing and style that form the bases for early professionalization and which can reinforce a sense of community-building within the classroom setting and beyond. This is particularly important as faculty members strive to have students submit their research papers to peer-reviewed journals that require an academic style for all submissions. Furthermore, by integrating bibliography and reference management tools, the Nota Bene suite also stimulates professional training through the use of Ibidem, which facilitates the practice of accumulating and organizing bibliographic resources in a controlled database structure. In addition, cognitive and organizational issues are at play as Ibidem stimulates the students' keyword imagination and functionality.

Finally, the Orbis application that forms part of the suite serves as the cornerstone for enhancing the course's cognitive objectives. By working in the background to index and manage the vocabulary used in documents, Orbis serves as a mirror, providing researchers–the note-takers–with their own terminology and argumentation structures. In other words, *how* note-takers conceptualize their readings is reflected, in turn, by Orbis through its search and retrieval services. Here, then, we can capitalize on the need to sharpen the students' conceptualization skills, their abilities to categorize data and structure hierarchies, and their comparative framework structures. The quality of the retrieved content depends on the sophistication of the search, which, in turn, is conditioned by the conceptual acumen brought to bear by the user at the data-entry stage. Finally, two elements of Orbis make it particularly useful and rewarding for the student-researcher. First, Orbis places no obligation on users to know anything regarding names of files or their locations on the computer's disk. Second, documents indexed by Orbis can be freely edited as frequently as needed. Orbis automatically updates documents, whether added or changed, and makes them readily available for additional searches and retrievals. This feature encourages students and researchers to fine-tune their notes (or any documents) as new data, ideas or references become available.

Support

Clearly, students in courses involving software applications need training and support, functions which, in this case, fall on me insofar as centralized support at the university is generally limited to a relatively narrow slice of widely used applications. Training sessions are mandatory components of the course, as stipulated in the syllabus. Students are required to attend four three-hour training sessions at the start of the semester, one session for each of the suite's applications, plus a summary session designed to reinforce previous acquired skills and focus on the integration of all applications. In addition, I hold virtual office hours (VOH) for technical support issues. These were normally scheduled during hours in which I was home and not in the office and could thus be at my computer working on other projects.

It was essential that the training sessions resonate with students' experiences and needs. Key to this was the use of the students' own notes–anything, actually, they had written–which they had used as a part of their preparation for other courses or for previous research projects. I did not begin the training sessions with the word-processing application, partly because I wanted to highlight that, on the whole, differences among such applications are not significant and that the learning curve was minimal. Instead, I concentrated on training the students in the use of Orbis for the purpose of extracting data that had already been entered, in this case, by one of them. Thus, the first exercise in the computer training laboratory entailed taking several of the students' files written with any application (invariably, it was Word) and having Nota Bene convert and save them. We then quickly built the Orbis database and were soon querying it for terms or combinations of terms. The first time we did this, the class underwent a collective moment of apprehending the essential value of conceptualization in ways that I could have not planned.

I asked the first student who volunteered to have his files converted and for whom we built a database of what the files contained. "These are files dealing with my project on inter-generation transmission of property," he said. The training session was now in earnest. "Fine," I said, "let's have you search for such terms." Once he entered the proper boolean-based search, the program responded that no such text could be found. The student insisted that the contents were as he had described them. I could sense that the students were now entering into the mode of discredit that sometimes accompanies "authority figures." As we entered more deeply into the conversation in front of the other students, we arrived at the point when he said, with some frustration, that the entire project that these files dealt with had to do with "dowries." At this point, I asked him to enter the term "dowry," and in an instant, several passages of the student's files were presented by Orbis for selection and incorporation into a document. The point had been made and the students' collective gasping sounds signified their understanding that *how* a process was conceived by the researcher makes a great deal of difference. The student *could have* conceptualized "dowries" as having different operational meanings, including "inter-generational property transfer," or "parental control," or "family economies," or any other ideas under which a discrete aspect of the researcher's project could be conceived. The students' gasp proved reassuring. The rest of the training would pose few other challenges. One of my central objectives, to make explicit the need to theorize, to rise above the many and complex details of data, had begun to be reified.

For the VOH, I have used AltME (Alternative Messaging Environment) from Safeworlds.com, which facilitates the setup of a chat room, using one of my computers as the server. Advantages to ALTMe are multiple, but in particular, it provided an environment that is immune from commercial interference and which eliminates concerns for privacy, server or network support, load or network traffic considerations, student registration databases, or locational limitations. The personal computer, the simplest unit in computer hierarchy, was the server, eliminating (in practice, if not in principle) technical or hardware support considerations. In fact, the connections never failed and the system ran "24/7."

Finally, I created a web site for the course. The site (www.fiu.edu/~szuchman/palmnotabene03) provided students with important components. These included the syllabus, several training documents, including the treatment and processing of WordSmith files on the Palm Pilot and synchronizing them with Nota Bene on the computer, contracts governing students' use of hardware and software prior to their distribution, needed links, and a subset of the reading assignments in PDF (readings were secure, access information was available only to registered students).

How did the class function? What activities were required of the students? How was technology employed?

Class Dynamics: Seize the Moment

First, students are required to bring their Palm Pilots and keyboards to every class meeting. This provided them with the tools to take notes "live" during class discussions. In addition, students are required to take notes on the weekly assigned readings using their Palm Pilots. As a student makes an in-class presentation on one the reading assignments, discussions ensue, everyone has the opportunity to add to their notes, adding new perspectives, data–and, in particular–theory and framework through which to analyze the data anew.

Indeed, it is the "live" or dynamic nature of their Palm-based WordSmith documents that serves two functions. First, it provides students with the ability to update their efforts. Free from the static quality of typical word-processed documents and the limitations they present, including inhibiting revisions, students in class are capable of dynamically updating their notes in step with the flow of the in-class discussions. Furthermore, once their notes were "refreshed" by the class discussions, students had opportunities, as the semester progressed, to update notes taken on previous weeks' assignments as indicated by the connections made during discussions that further developed previous issues. Students would synchronize their WordSmith documents with their computers. Whenever they opened Nota Bene, Orbis would index and/or re-index the notes, keeping all such documents live and dynamically linked.

Second, the very image of the Palm Pilots and keyboards in front of the students sitting around the seminar table served to reinforce in me, as instructor and discussion steward, the need to "seize the moment," that is, to bring out and emphasize concepts, categories, and relationships arising from the materials and the discussions. And beyond the classroom setting, to provide students with the tools to process the material, aided by their digital media and the software that would provide for such concepts and relationships well the past the end of the course. In this fashion, both my students and I have changed our own learning and teaching paradigms in ways that I believe lead to superior results and to a developmental approach. At the end of each class meeting, students beamed their notes from their Palm Pilots to mine. I found that, in addition, students exchanged notes with each other, surely evidence of collaboration and community building (and evocative of Star Trek conventioneers).

What is the purpose of having students beam their notes to my Palm Pilot? After accumulating the students' notes on a weekly basis, I transferred them to Nota Bene on my laptop and used Orbis to index their contents. In other words, I created an Orbis database of the students' notes, which provided me with the opportunity to review and assess the nature and scope of the students' understanding of the course as reflected in their files. I was able to assess by individual student(s), by reading assignment(s), by author(s), or by theme(s). The capability to identify data by student, date, author, or terms powerfully extends instructional opportunities. I reinforced best practices by displaying on screen through an LCD projector the results of Orbis queries on my computer during class periods set aside for communal reviews. These sessions are designed to critique and strengthen research annotations within a friendly and supportive environment.

Faculty and students participate in this process. We look for ways to embed summary concepts represented by the student-entered data. Two objectives are addressed. First, we engage in an ongoing dialog dealing with the essentials of authors' arguments, for example, the place of such arguments within the framework of theoretical and conceptual categories, and/or the schools of thought where authors and arguments belong within the broader literature. Second, I illustrate how to embed the ideas and concepts that we consider appropriate during theses sessions in order to summarize relevant sections of students' notes, yet without cluttering their readability. I show them how notes taken, for example, on cases of provincial resistance to national authority in the nineteenth century would pop up searching in Orbis for terms such as "militarism," "nation-building," "regionalism," "integration," or any concepts that resonate with the researcher's data. Because Orbis is a free-form database, students can place these non-printing items at will anywhere in their documents. This is a good example of how the use of technologies dedicated to the discrete but ample audiences of learners and researchers can serve to enhance the meta-cognitive process, which in this case represents an essential component of the course design (Flavell, 1979).

Sometimes, students provided the best validations of the objectives designed into the course. For example, I make the point during the first class meeting that retaining the vital information in the vast amount of reading materials we encounter cannot be effectively accomplished without the use of additional helpful devices. I then take a poll, asking how many of them use underlining, highlighting or bracketing of portions of the texts they read. They all do. When asked why they do these things, invariably they assert that they help to retain the material. I then challenge them with the evidence that their efforts do not yield the hoped-for results. I proceed to explain how, by contrast, the technology they will learn in the course will provide the tools for recovering and accessing data. One student mentioned in class at roughly the mid-point in the semester that he had experienced a computer glitch and had lost the file that contained his notes dealing with one of the readings that had been assigned only six weeks earlier. He remarked to us how re-reading of the assignment felt like a completely new reading, and that he had only a modest recollection of the author's essential arguments and empirical evidence. In noting how surprised he was to "rediscover" the material despite its recency–and commenting on his new doubts about retaining it—he remarked on his heightened appreciation for the new methods and technological skills he was learning in class to address these matters. It represented a classic "learning moment."

Students were required to write a literature review approved by me on a topic associated with the theme of the seminar. As part of the planning discussions that I held with each student, I required the submission of preliminary bibliographies. This exercise served, first, as practice involving skills in data entry into the bibliographic management database, and, second, as the basis for discussing the content, organization and conceptualization of the paper. Papers were submitted as electronic attachments. This provided me with the opportunity to observe the skills each student brought to the use of the software. In general, students appear to make good use of Ibidem, to the extent that their references are both richer–more so than in papers without these software tools–and appropriately cited, as required by the Chicago Manual of Style. Although the quality and strength of argumentation are conditioned by each student's intellectual abilities, arguments appear to be presented more coherently, showing fewer digressions from the papers' stated objectives. It appears that Nota Bene's emphasis on the logic and benefits of heads and subheads, as the framework surrounding research-based writing, has had positive effects on the students' organizational skills.

Students' Feedback

The feedback given by students to this mode of delivery in the form of course evaluations has generally yielded unusually great amounts of written responses. As is known about most advanced-level courses containing modest numbers of students, the numerical tabulations generally indicate high levels of course and instructor satisfaction. However, in addition to ranking course and instructor on a conventional 5-point scale, students have shown an unusual interest in writing comments about their experiences in this type of course, and they, too, have been generally positive. I take this amount of voluntary written comments as a significant response insofar as they are generally rare and, when offered, parsimonious. I built on this unusual sharing of views and requested the students' anonymous feedback on specific dimensions prior to the end of the semester and independently of the course evaluation process.

Students were asked to comment freely on the following issues:

- the integration of content with technological instruments

- the communal aspects of the technological learning processes
- the communal aspects of the conceptualization processes
- how the nature and the "architecture" of the course compare with other courses in the context of connecting with the history subject mater
- limitations
- other relevant matters

They were given one week to submit their responses in sealed, unmarked envelopes provided by the Department of History. In their feedback, students reported initial apprehension about the combined intellectual and technical challenges, on the one hand, with a significant level of satisfaction in meeting them, on the other. They especially valued the software training workshops but had little to say about the electronic chat room that was made available to them for their continuously-available group communications, in addition to on-line weekly meetings with me for technical support. In fact, very few would present themselves for these virtual office hours, and the few who participated did so only infrequently. They reported very favorably on the portions of class sessions during which their notes, displayed on screen, formed the subject of discussions designed to improve the conceptualization of their contents and the mechanisms for entering those concepts into their Orbis databases. Here is where they linked the considerable amount of assigned reading, commenting on the lasting benefits of the mode of digital note-taking.

Every respondent reported on the unusual nature of the course design and the adjustments needed to incorporate the matrix of ingredients. All of them commented that they had never experienced such a combination of elements. But they also reported wishing that they had had this experience earlier in their student careers. Some suggested that the combination of course content with the technology should serve as a template for the department's other courses.

One respondent pointed to the text-oriented nature of the content as a limitation, noting that the Ibidem and Orbis applications were not obviously suited to multimedia data. Finally, they registered a sense of hopeful–even confident–expectation that they would be building their personal databases both of their own electronic bibliographic libraries and of their annotated empirical qualitative data. A few registered their hopes that these tools and data would prepare them for additional graduate work, including master's theses and doctoral dissertations.

What have I learned from these experiences and what would I do differently? First, I have been reinforced in my assessment that one cannot underestimate the learning capacity of even jaded graduate students, even in matters that may initially frighten them. But I have also learned not to underestimate how much reinforcement is needed in the realms of technology learning, use and retention. Implementation of software features that would appear obvious to me was not always obvious to a few students, and I discovered that we would course through the same technological material more than once. Ironically, I found that students do not instinctively take notes on what they observe on screen during training sessions.

In this regard, I have already begun to address these matters. First, I will eliminate the chat room and virtual office hours. Students have not generally taken advantage of these facilities. They may well be too embarrassed to share their technological difficulties or they may be frustrated by the medium's limited capacity to express their technical needs.

Second, I will be creating computer training videos, short exercises based on capturing screen activities involved in different software tasks. The videos will be accompanied by the sound track of my voice highlighting the processes observed on-screen. These videos will serve to reinforce basic information covered in the initial software training workshops in the computer lab. The videos will also include tips designed to increase the students' confidence levels. These reinforcement videos will be distributed on CD's so that the students can use them at will.

Summary and Conclusions

Fields based on qualitative analysis require significant data entry efforts on the part of the researcher. The Palm Pilots' modest cost and wide array of applications provide ultra-portability and ease of use, enhancing student-researcher productivity and idea-sharing. The portable keyboard makes the Palm Pilot a viable and cost-effective alternative to laptop computing for data entry, particularly in remote sites such as archives, libraries, and, of course, the classroom.

Desktop applications suitable to critical-thinking tasks and designed for the academic enterprise provide tools for qualitative data analysis and offer extraordinary potential, as in the case of Nota Bene. The tasks involved in expanding the cognitive learning processes and providing the faculty with opportunities to address students' meta-cognitive objectives strongly reflect the need that technologies which support the learning endeavor must be encouraged. At advanced levels of study, additional benefits are found in applications that accelerate students' awareness of their disciplines' professional conventions, the protocols involved in written work, and the tools to implement best research, analytical, and writing practices.

Footnotes

[1] See the *Who Built America?* project centered at the City University of New York at http://www.ashp.cuny.edu/index.html. For history in the United Kingdom, see http://www.gla.ac.uk/~histtltp/tutor.htm from the University of Glasgow as a good example of a rich site. Another exemplary Web resource (http://www.gla.ac.uk/~histtltp/tutor.htm) in North American history is the Valley of the Shadow Project at the University of Virginia.

[2] See (Wittgenstein, 1961, section 43) for discussion on the meaning of words and an emphasis on context.

[3] My use of the term "sea change" may be considered hyperbolic by some, but the Palm keyboard received PC Magazine's "Product of the Year" award for its technological innovation and design.

References

Case, D. O. (1991). Conceptual organization and retrieval of text by historians: The role of memory and metaphor. *Journal of the American Society for Information Science*, *42*(9), 657–668.

Fink, L. (2001, February). New tidings for history education, or lessons we should have learned by now. *The History Teacher*, *34*(2), 235–242.

Flavell, J. H. (1979). Metacognition and cognitive mentoring: A new era of cognitive developmental inquiry. *American Psychologist*, *34*, 906–911.

Palmer, S. E. (1975). Visual perception and world knowledge: Notes on a model of sensory cognitive interaction. In D. A. Norman, D. E.

Rumelhart & LRN Research Group, Explorations in Cognition. San Francisco: Freeman.

Preston, G. (1996). Seminars for active learning. In A. Booth & P. Hyland (Eds.), History in Higher Education: New Directions in Teaching and Learning (pp. 111–127). Oxford: Blackwell Publishers.

Steiner, G. (1996). Learning: Nineteen scenarios from everyday life. (J. A. Smith, Trans.). Cambridge: Cambridge University Press.

Wineburg, S. (2001). Historical thinking and other unnatural acts: Charting the future of teaching the past. Chicago: Temple University Press.

Wittgenstein, L. (1961). Tractatus logico-philosophicus. (D. F. Pears & B. F. McGuinness). London: Routledge and Kegan Paul. (Original work published 1922).

Project CineMath
An Adventure of Video Signals

George Kafkoulis
Florida International University

Abstract

The author's own experiences as a graduate student, when he took mathematics classes that were taught the traditional way, instilled in him a desire to change and improve the way mathematics is taught at the university level. As a faculty member in a university he now integrates technology into his teaching. Using simple and cost effective tools, he creates and archives movies of the lecture event, including audio and notes that are handwritten using a digital tablets. These movie files, which are quite small in size are stored on the web and can be easily retrieved and viewed as many times as necessary by students. The author describes his experiences and responds to some possible concerns about the technology that he uses in his classes.

Introduction

When I was a graduate student, most of the math classes that I took were taught in a traditional manner. Ever since that time, I have felt the need to improve the way that mathematics is taught at the university level. I did learn math in the classes that were taught using traditional methods. However, nowadays many students come to the university without much preparation in mathematics, and they find it difficult to learn mathematics when it is taught the traditional way. Many math faculty members at universities across this country and abroad are now using technology to help them teach mathematics. In this chapter I will describe my own attempts at integrating technology in my calculus classes.

My Early Efforts At Integrating Technology

During the academic year 1992-1993, I became involved in a project with my colleagues titled "Instructional Mathematical Laboratory[1]" that tried to enhance the content of our Calculus lectures. We used software-based notebooks displayed on the computer monitors and capable of displaying graphs and of making complex calculations. Even then, I tried to balance those sessions with a heavy presentation of traditionally taught topics, simply because the monitors were not capable of displaying some of the fundamental mathematical notation. During the fifth year of the project, I began to decrease the use of the electronic notebooks and started paying more attention to the mechanics of the mathematics lecture. I came to the conclusion that although the beautiful examples that you can build with software tools such as Mathematica & Maple are motivating and inspiring to the Calculus novice, they cannot provide the student with the necessary insight to use Calculus itself as a tool. The practitioner of calculus has to understand much more Applied Mathematics in order to use Calculus as a tool. Even the software tools that provide mathematical pictures and complex calculations require the user to have more mathematical insight.

My Model of Teaching Mathematics

Based on these experiences, I followed teachings styles that inform and inspire the student to enter much deeper into the Mathematical field than the average classroom encourages. I changed my teaching style based on the premise that teaching fewer black-box techniques will shift the focus of the student from learning just techniques to understanding the key ideas as well as the techniques. There is no denying the fact that the Mathematics Professor is a facilitator of learning. I am not so certain that the learner can discover by herself mathematical phenomena that do not have analogies in the real world such as the density of irrational and rational numbers, densely perforated graphs, now-where monotonic functions, and non-integrable functions. All of the above mentioned phenomena are always a few lines away from the standard examples that we usually teach such as continuity of a function at a point, Riemann integration, positive derivatives at a point etc. What may seem as pathologies are in most cases the generic behavior of the "majority" of the mathematical objects as theorems of real analysis suggest (in the context of sets of second category in families of functions). Thus, not "going below the surface" in our lectures may be also pedagogically inappropriate and intellectually dishonest.

I advocate a teaching model that raises the role of the Mathematics Professor from that of a facilitator to that of an intellectual authority who will guide the student through the jungle of mathematical ideas, logical arguments, mental pictures, patterns, calculations and theoretical tools. This of course does not contradict any variation of this teaching role to that of facilitator or advisor in an adventure of students' self-discovery of carefully pre-selected mathematical content. Nor does it contradict any efforts to reform the content and use technology to investigate simple phenomena that are amenable to such presentation. In my teaching models, such variations require careful considerations and play a secondary and auxiliary role.

Based on this model of the role of the Mathematics Professor, I examined very carefully the activity of the teacher and the students during the lecture. In a traditionally delivered lecture the teacher handwrites sentences on the board about mathematical objects and the justification of their validity. These proofs, explanations, and computations are supplemented by oral remarks that draw from the professor's scientific background, research, point of view of mathematics and other experiences. A good mathematical lecture is a partially orchestrated and partially spontaneous (as unexpected questions by the audience forces the professor to go out of the script) event. As such, the "lecture-event" needs to be somehow recorded and archived.

Traditionally, students try to capture a professor's lecture-event by note taking, especially when the professor presents topics and proofs that do not appear verbatim in the textbook prescribed for the class. In the attempt to take accurate notes, students may not fully understand the ideas presented and the reasoning behind them. Even assuming that accurate note taking is possible, the mathematics lecture, with its dynamic characteristics and its deductive flow of ideas cannot be completely and accurately reproduced from handwritten notes, As a consequence, unless the student is truly motivated to study mathematics, she will have difficulty studying effectively and efficiently outside of the lecture hall. Although, self-discovery through books is possible, the first initiation to a new mathematics concept takes place during the lecture. Of all the parallel discussions and comments, it is this experience that students usually try to capture by note taking and simple memorization. Do they succeed? I think not.

Over the years, my efforts to improve learning in mathematics concentrated on mathematical content. I now believe that of equal importance is the delivery and capture of the lecture and students' study methodology, tools and habits. Thus, I began by reevaluating note taking, since it is a key student activity during lectures.

Painting and Handwriting

Note taking undoubtedly has pedagogical value. However, during the first explanation of the ideas (i.e., during a lecture) understanding is far more important than keeping a record of the presented knowledge. I decided to keep note taking as an out-of-classroom activity and an integral part of the students' studying methodology. Instead of writing on a chalkboard, I now write on a digitizing tablet that is connected to a computer whose screen is projected on a screen or the wall by a Liquid Crystal Display (LCD) projector. I use a good painting program to handwrite (paint a jpeg file), create color overlays of highlighting and movable layers for low cost and easily produced animations.

During my lectures, students are encouraged to try to understand the presentation and not to take notes. Periodically I save the picture file of the painting program. As soon as the instructor saves the working picture file, it is displayed on the monitors of the students' workstations by the browser that is installed in the computers. A series of web pages organizes the content by lecture and by subject. The student can browse through the lecture notes written a few minutes or a few days ago. By writing or painting on pre-assigned files on a web server that is accessed by the students' web browsers in the lecture hall, I am able to accomplish a low budget and primitive broadcast of the contents of the "electronic chalkboard" to the students' workstations.

In practice and on his or her own initiative, the student browses through the previous notes in an effort to understand the current concept, either by revisiting a forgotten definition, or an analogous argument. These efforts started during the Fall 1997 semester in the NextStep PC lab in the Mathematics Department at Florida International University (FIU). Although students had nicely printed accurate notes and their attention was dedicated to understanding the oral and written arguments, the lecture-event still was not being adequately captured for further study and review.

The "Animated Notes" Paradigm

During the Spring 2000 semester, I submitted a proposal to FIU's iNET initiative, which is a campus wide initiative to support and encourage faculty members to integrate technology into their teaching. The title of the proposal was "The digitizing of a mathematical lecture." It was during the first month of iNet training that I got the idea of capturing the handwriting and all other teaching related activities on the computer screen in a low cost movie format. Screen capturing software existed at that time, but the combined use of the handwriting using the tablet (I used a Wacom USB tablet) and the CPU load necessary for the computer to capture its own screen in a movie turned out to be a daunting task even for the fastest computer that was available to me at that time (and perhaps even today!). During a meeting with an instructional technologist who was and playing with converters and two computers, we realized that a second computer could do the "dirty job" of capturing the screen of the first computer. That led me to a frantic exploration of the world of codecs, and other video and multimedia gadgets.

My explorations revealed that it was possible to do what some of the experts on video at the University were explaining in long e-mails as impossible. I had to learn about sound-input at a guitar store and had to test and return several gadgets from a computer store. The first movie that I produced was low in resolution (320x240) and USB-captured using a program by Real Networks called Real Producer. I started teaching a course with it in September of 2000. The Real Networks product was giving me a hard time editing the movie and the small size of the movie window was not satisfactory. So I began experimenting with a Firewire port (that I was hoping could allow me to capture more information) and Macintosh G4s. I used Sorenson Broadcaster on a 400Mhz PowerMac G4 but the quality was still poor at large movie window sizes. Finally in October 2000, I managed to use Sorenson Broadcaster to capture with Firewire (in a way not intended by the software) a movie of size 768x480 with surprisingly good quality for handwriting and in a very small file size (25MB for 75 minutes) which I was streaming live over the internet (modem band-width) and archiving in QuickTime movies that could be cheaply edited at will. This was the technology that I named "Cine-Math[2]."

CineMath became a system that captures the handwritten lecture notes, the audio and other computer screen activity such as graphic animations in a movie format that is streamed live to the World Wide Web via the internet and is recorded for streaming on demand in high resolution (768x480). CineMath is the product of an "engineering" effort to combine several computers, redirect their video and audio signals through a Firewire port to a broadcasting software that produces streaming movies accessible through a system of web pages that incorporate Javascript and DHTML layers and Macromedia Flash.

At the end of the lecture, the teacher has to use the encoding computer in order to transform each lecture movie to a streaming movie file and transfer the files to a streaming server. We refer to these movie files as "CineMath files." Since there are no video cameras (& no filming crew) involved for making the CineMath files the size of the files is very small. It takes about five to ten minutes for post-processing of the files (the process is fully automated through AppleScript, Unix Shell scripting and awk), and the resulting file sizes are approximately 40 MB for a one hundred minute Calculus lecture. In addition, if a web cam is used in order to capture the teacher's body language and gestures, then the file size (of the webcam file) is approximately 200 MB at half the resolution of a CineMath file.

Reusable Teaching Objects

Using CineMath and the web, each lecture is archived into two forms. The snapshot of each lecture-page appears on the course web page. These notes usually become an important supplement to the textbook. Students are required to rewrite the notes and combine them with the textbook to develop a set of comprehensive notes that will be used by them for reviewing the content and preparing for the exams. During this critical rewriting period, students revisit the lecture, which is accessible on the web as a streaming media file. As an additional incentive for rewriting the notes, the professor places questions in strategic places of the lecture notes and requires students to find the answers. For most of these questions, students have to replay the CineMath media file using the QuickTime media player. The professor may also have prepared and archived mini-lectures or tutorials on additional topics or on topics not covered extensively in prerequisite courses. These CineMath files are the "reusable teaching objects." They carry the audio-visual content of a mathematical lecture. Students can replay and view these media files in full-screen resolution of 768x480. Although the used encoding is designed for 56k modem streaming, in practice a faster connection is needed. However, the size of the files for each lecture is so small, that students can download them into zip disks and view them offline. In fact the movie files for a whole semester of Calculus lectures, which average about seventy-five minutes each, could easily fit in a single CD-ROM.

The "Cinematographic" Electronic Notebook

The collection of CineMath lectures or tutorials would become notebooks that, when opened, are handwritten in front of the reader's eyes at the time of reading, while the author who is the professor, narrates the unraveling content. The technological solution CineMath offers an efficient, inexpensive and highly automated method for the creation of theses books. It is not within the scope of this chapter to propose a theory of mathematical reading and understanding. However, there is no doubt that static text presented on paper or a computer screen is unable to unravel the thoughts and justifications that usually follow the writing during a lecture or exposition by a professor. On the other hand, "cinematographic" electronic notebooks function in a different way than the static books that are presented on paper or computer monitor. I do not know if this dynamically presented content is essentially related to a "Mathematical Language Processing" which can be considered the mathematical analog to "Natural Language Processing." I have found no literature on this subject. This project timidly suggests that we should study the effectiveness of this dynamic presentation of mathematics text and pictures on learning outcomes.

A Pilot Program

Personal communications from students indicate that they appreciate the opportunity to relive the lecture event. Although I am very much excited about the capabilities of these new technologies and techniques and student reception is very good, no independent evaluation or comparison studies have been performed for this pilot program. During several semesters, I provided students with CDs every two weeks in order to compensate for the inadequacies of the school computer multimedia infrastructure and the inequities associated with students' access to computers outside the university. Without a doubt, improvements in the hardware at the university and at home as well as the availability of more bandwidth will give CineMath the opportunity to be more widely available. The students have embraced the technology, something that is apparent from the comments made on the end of the semester student evaluations of teaching.

Some Concerns and Responses

Concern: The students become passive listeners or viewers instead of being active note takers.

The professor has to interrupt the lecture frequently by asking questions, forcing the students to go back-and-forth through the lectures using the web browser or other software tools. In addition, the teacher purposefully asks the student to work on paper for a few questions that relate ideas from previous pages or previous lectures. These tasks are scattered throughout the lecture and they make sure that the students are active and alert.

Concern: The proposed reforms will eliminate office hours and personal contact between students and the teacher.

All the traditional classroom based teacher-student activities remain intact. The quality of students' questions will improve with the use of this technology. Since the students will have the opportunity to replay the entire lecture they will be able to find answers to obvious questions.

Concern: I fear that the calculus course taught by Professor X will be replaced, by the streaming movie files or the CDs of calculus taught by Professor Y.

Without a doubt, the availability of Professor Y's CineMath lectures may create a noble competition between professors. This alone may have a positive influence by forcing Professor X to revitalize, rethink and renew stale lectures, presentation habits or blackboard etiquette.

Conclusion

There are technological ways to password-protect streaming files. This may provide a certain degree of privacy for a particular course. However, if this project's evaluation provides evidence for better learning outcomes, then traditional lecturers will have a hard time justifying the rejection of the proposed enhancements. This project may even lead to the creation of an archive of several different teaching styles and techniques on the same subject, which could be an important contribution to the diversity and richness of the academic environment.

Footnotes

[1] A project with co-PIs Gerardo Aladro and Enrique Villamor of FIU, funded by the NSA-program for Minority Institutions.

[2] The address for the CineMath web site is: http://dmcalcue2.fiu.edu/~kafkouli

Human Patient Simulators in Medical Students' Preclinical Curriculum

John L. Szarek
Ian Winston
Ross University

Abstract

Human patient simulators are becoming increasingly common in the education and training of physicians in the clinical portions of their curriculum, however their use in the preclinical portion of their training is only beginning. In this chapter we describe how we incorporated the human patient simulator (HPS) into the preclinical curriculum at Ross University School of Medicine. The impetus for developing a simulator program arose from our desire to incorporate the HPS into the students' last preclinical semester as an evolution of our existing problem-based learning (PBL) program and to prepare students for their first clinical experience. We articulate the method we devised for case development, and the design of and progression through the simulator PBL sessions. We share the responses of our students to the program that delineate several significant differences between our traditional PBL sessions and the simulator PBL sessions. Challenges to implementation, including the issue of whether learning is enhanced, cost, faculty contact time, and student group sizes are discussed in the chapter together with potential solutions. Based on our experience we found that faculty members who are champions of this method of learning facilitated incorporation of the human patient simulators into the curriculum and that using a HPS enhances the benefits of PBL in terms of students' motivation to learn, interpersonal skills, and application of knowledge. Finally, we offer five take-home points for educators who may be considering implementing a similar program at their institutions.

Introduction

Simulators are well-established technology for training personnel involved in disciplines such as aviation, the military, nuclear power plant operations, and business (Issenberg, McGaghie, Hart, Mayer, Felner, Petrusa, et al. 1999). Over the past few years human patient simulator (HPS) technology has experienced an exponential growth worldwide. In 1994 there were less than 10 centers worldwide, now there are more than 300 centers (McIndoe & Jones, 2003). Simulators have been used in anesthesiology training for 30 years and, as advances in technology led to the development of more sophisticated simulators, they have been used in educating medical students in their clinical clerkships and in their residencies (Issenberg et al., 1999). Their use has only recently been incorporated into the preclinical education of medical students.

The traditional undergraduate medical education structure consists of two years of basic science courses (preclinical) plus two years of clinical rotations. During the preclinical curriculum students learn the science basic to medicine in such disciplines as anatomy, biochemistry, physiology, and pharmacology. Traditionally, these were taught as separate courses in medical schools but a majority of schools are making changes in their undergraduate curricula to meet the changing needs of society and medicine (Milbank Memorial Fund, 2000). Some of the curriculum reforms have been directed toward pedagogical methods whereas others adopted approaches to foster integration of content across the sciences and integration of basic science and clinical content. During their clinical years, the students spend several weeks rotating through various clerkships including internal medicine, family medicine, pediatrics, surgery, and gynecology/obstetrics. It is during the clinical years that students receive their first intensive education in clinical practice.

In a survey of 60 simulation centers which reported using simulators in the education of medical students, anesthesiology residents, nurses, paramedics, respiratory therapists, or army personnel, about half reported using the human patient simulator for physiology and pharmacology in the undergraduate curriculum (Morgan & Cleave-Hogg, 2002). Although the proportion of the centers that used simulation technology for medical student education were not delineated in the study, there have been a few reports describing the use of human patient simulators in the education of medical students in their preclinical years (Euliano, 2000; Euliano, 2001; and Tan, Ti, Suresh, Ho, & Lee, 2002). In this chapter we will detail our experience in the development of a simulator program in our preclinical medical curriculum at Ross University School of Medicine (RUSM) in Dominica. We chose not to provide an exhaustive review of the literature on their use, however, we included pertinent references that were relevant to our discussion.

The Simulator

In order for our students to learn in an environment similar to that which they will encounter when they begin their clinical rotations, our human patient simulator laboratory was set up to resemble a clinical examination room complete with a life-size mannequin and its computer interface (HPS Version 6, Medical Education Technologies, Inc. [METI], Sarasota FL), and a patient monitor with a six-channel display (Viridia 24/26 Component Monitoring System, Philips Medical Systems, Sunrise FL). The mannequin has many attributes of human anatomy and physiology. The respiratory system includes a realistic airway that can be instrumented. The mannequin breathes in room air (either spontaneously or by a ventilator) and exhales carbon dioxide, and chest excursions occur with breathing. Breath sounds can be auscultated with a stethoscope. The mannequin responds appropriately to oxygen therapy and, conversely, will suffocate and die if a plastic bag is placed over its head. Features of the cardiovascular system include palpable radial, brachial, femoral, and carotid pulses that can differ in quality according to the clinical situation. Heart sounds can be auscultated with a stethoscope and can have different characteristics based on the clinical situation. The mannequin's eyes blink and its pupils respond to light. It has interchangeable male and female external genitalia that can be catheterized to quantitate urine production. There is a speaker in the mannequin's head that allows it to talk by means of an actor speaking into a wireless microphone from a control room. More than 50 drugs can be administered through an intravenous stopcock. The drug's identity, concentration, and dose administered are automatically detected by means of a bar code reader next to the stopcock. Sophisticated algorithms accurately depict the pharmacokinetic and pharmacodynamic effects of the administered drug. As will be described later, scenarios can be developed replicating human clinical scenarios. The simulator can be controlled by a trained facilitator from the bedside using a remote computer linked via a wireless connection to the main computer or from a remote control booth.

Implementation

Impetus for Incorporating Simulation at RUSM

Problem-based learning (PBL) originated at McMaster University Medical School in Canada, where, in the late 1960s, PBL was adopted as its major teaching/learning approach. Several other schools followed in the 1970s and since then PBL has become increasingly common in medical schools and curricula (Schwartz, Mennin, & Webb, 2001). Although the term PBL may have different meanings among medical schools, Howard Barrows, one of the originators of PBL, described the attributes of a true program (Schwartz et al., 2001). In a typical PBL session, students work in small groups with a faculty member who is a facilitator of discussion and learning rather than a resource. In the first session, students encounter a problem cold without any preparatory study in the area. The students interact with each other exploring their existing knowledge as it relates to the problem. They form and test hypotheses about the mechanisms that might account for the problem. They identify further learning needs in order to make progress with the problem. They undertake self-study between group sessions to research the learning needs and upon returning to the group, they integrate their new knowledge and apply it to the problem. New hypotheses are formed and tested, and the process is repeated as needed. Finally, the students reflect on the process and on the content that was learned (Schwartz et al., 2001).

During a typical session upon receiving a new case or new information, the students record what information they have about the patient and the patient's problem (data), they hypothesize causes of the problem (hypotheses), they explain reasons for their choices of hypotheses and decide what information they need to help them prioritize their hypotheses (analysis), and then the students determine what other concepts or areas of knowledge they need in order to explain a hypothesis or some other aspect of the case (learning issues). The data, hypotheses, analysis and learning issues are recorded on flipcharts. After each session and after completing a case, the students and facilitator summarize the content that was learned and reflect on the process. Our PBL sessions are conducted at RUSM as described above over three 2-hour sessions with a week between sessions for students to research their learning issues. The program is conducted in this manner during the students' first three semesters. Each semester has two directors of the PBL program and the directors collectively form a subcommittee to the curriculum committee. In order to continue the PBL program in the students' fourth semester and to achieve the goal of preparing students for their first clinical rotation, we decided to incorporate the simulator into the fourth semester curriculum and to conduct the simulator sessions in the same manner that students were already accustomed to in their PBL sessions.

Simulator PBL Case Development

In the development of our first case, we decided to use an existing paper case from our third semester PBL program. Using an existing case from our established PBL program facilitated rapid deployment of the simulator into our curriculum in the fourth semester. The case had been included in the curriculum for three semesters and thus had been tested. We did make some minor changes to the case including the age of the patient and the circumstances surrounding "his" need to visit the clinic.

Regardless of whether existing cases are used or new cases developed for the simulator, the first step in developing a case is defining the learning objectives. For our cases at RUSM, these are determined by the case writers, in conjunction with the course directors for the courses in which the students are enrolled concurrently with the PBL program. For instance, in our initial case of an acute asthmatic attack in the emergency room, objectives included such concepts as the common causes of wheezing, the pathophysiological processes involved in asthma, drugs used to relieve an acute asthmatic attack and the mechanism of action of the drugs used in the long term control of asthma. We next constructed the case to ensure that the students would be able to achieve the objectives after completing the simulator sessions and the patient's history and pertinent physical examination findings were described. For example, the case that we used for our asthma simulation follows:

> Steve Austin is a 23-year-old male who was at work as a painter/decorator when he became short of breath. He was painting indoors at the time. His breathlessness became worse and he had to stop work and get some fresh air. Eventually his boss told him to go home. On the way home his symptoms worsened so he brought himself to the emergency room instead.
>
> **Pertinent physical examination findings included**: General, a large man awake and sitting upright on a hospital bed, he is cyanotic with

audible wheezes, he is unable to speak in complete sentences and is very short tempered.

Past medical history: some hay fever but nothing like this before.

Family history: little brother with asthma.

Vital signs: Heart rate, 100 beats/minute (normal 55-90); Blood pressure, 128/76 mm Hg (normal 121/70); Respiratory rate, 32 breaths/minute (normal 14-18); Oxygen saturation, 88% (normal >95%); Peak expiratory flow, 330 Liters/minute (normal >600); Weight 68 Kg; Temperature, 98.6°F (normal 97.8-99.1).

There are limits to what the simulator can do and one of those is change skin color in response to hypoxia. Therefore, in order to emulate the physical examination findings associated with central cyanosis, we created small labels with the word, blue, written on them. These were placed under the tongue, on a fingernail and on a toenail, areas where students should look for signs of cyanosis when doing an examination of the respiratory system.

With the simulator, a case is represented by a number of scenarios. Each scenario is made up of several states that correspond to the physiological events occurring during the case. We outlined three scenarios for this example. In the first scenario, the baseline physiologic parameters were defined as above when the patient first presented himself in the emergency room. The next scenario was a period of improvement that the simulator experienced after appropriate drug therapy. In the last scenario the patient's condition worsened to the point where he exhibited the signs, symptoms, and physical findings as in the first scenario. We found it useful to have a clinician and a basic scientist involved in creating the cases. The clinician possessed the experience needed in order to create a patient with signs and symptoms similar to those that the students would likely encounter in a patient in their clinical rotations. Since part of their education is in physical examination skills, the presence of a clinician in the session assured that the appropriate physical examination techniques were demonstrated as students generated their data. The basic scientist, in our case, contributed expertise on drug treatment and the desired depth of understanding that students should demonstrate with respect to the basic science concepts.

Design of the Simulator Sessions

After creating our simulator case, we conducted our first simulator sessions with a group of students who had completed the four semesters and were in their first clerkship rotation. Our PBL program was not part of the curriculum for this group of students so they were not as adept at the process as our subsequent groups. However, what we learned from this group was that, as facilitators of the process, we needed to be sure that we incorporated the mannequin. Although the process parallels that in our paper cases in the typical PBL session, there were differences encountered when using the lifelike simulator. Mostly, these had to do with setting up the simulator and the clinic room, encouraging students to treat the simulator as a patient who needs their help, and providing data when requested by the students that usually was provided in the pages of the case.

The following semester we asked for volunteers from the student body in the fourth semester. We had enough volunteers so that we could run four groups of 10 students each through three simulator sessions. Initially students worked with the simulator in three one and one-half hour sessions held over a one-week period. Based on student feedback during debriefing sessions, we subsequently went to a trial in which students conducted the three sessions in one day. Each session with the simulator ran for one hour followed by one and one-half to two hours between sessions to research learning issues. We reasoned that this was more realistic in terms of how students will be working with patients in their clerkships. In our debriefing after each case we asked the students what they thought of doing three sessions in one day. There was general agreement that, because of the commitments the students face in their fourth semester, they appreciated not having extra homework outside the simulator sessions. A significant number of students did think that their experience would be enhanced if the sessions were spread over more than one day. As a compromise we decided to have students work with the simulator in two sessions on one day (Tuesday in the fourth semester), each separated by three to four hours for research. The last session would be held on either the following Thursday or Friday depending on the students' physical diagnosis laboratory schedule. We will continue to evaluate the timing of the sessions during the debriefing of each group of students as our program evolves.

In a typical PBL session the students encounter a problem "cold" without any preparatory study in the area. Since it was necessary for students to have had instruction on physical diagnosis so that they could generate the data for the case (for example, vital signs and other physical examination findings), we had to schedule the simulator sessions after the students had completed the appropriate physical diagnosis laboratory. Our curriculum in the fourth semester is based on organ systems. Thus the students not only had received instruction on physical examination skills, they also learned the pathology, microbiology, and pharmacology related to the organ system. Since the students worked with the simulator case after having completed the cardiopulmonary block, the simulator PBL case deviated slightly from true PBL in that the students had prior study in the area. However, the students did come into the case unprepared with respect to whether the case would involve the pulmonary or cardiovascular systems. A potential scheduling problem may arise as we develop additional cases to incorporate into the fourth semester since it will be necessary to conduct the simulator sessions after an organ system block is completed. For instance, it would probably not be possible to conduct a simulator session on an organ system that is scheduled near the end of the semester. Our intent is to provide students with a context in which they can practice applying their knowledge in preparation for their clinical rotations. Thus, it will not be necessary for students to be exposed to a case after every organ system block. We plan to focus our cases on those clinical conditions which are most prevalent in society.

Progressing Through a Session

The HPS sessions are designed to promote the "willful suspension of disbelief." We found that a short preamble describing the mannequin's limitations and how they should interact with it more quickly facilitated the suspension of disbelief. Without this preamble we found that students at the beginning of the session were more likely to look at the facilitator in the room for answers to questions rather than waiting for a response from the simulator. With or without the preamble, the students eventually gathered closer to the simulator, spoke directly to it, and treated it as a real patient. This was evidenced by students' comments during debriefing sessions in which they described their desire to care for and treat "the patient."

The sessions progressed at a slower pace than required in a clinical setting. In our typical PBL sessions, students were provided portions of the case on paper as they worked through the process. In the simulator sessions, the students were not given the patient history, physical examination findings or test results. Rather, the students were expected to use their communication skills to obtain the patient's history and physical examination skills to assess physical signs. If the students requested appropriate tests, the results were provided. Using the speaker built into the mannequin's head, the patient responded to student questions pertaining to his history and they also acquired the pertinent physical examination findings through the use of the skills learned in their physical diagnosis laboratory, for example, wheezing heard on auscultation of the chest. During the session we took advantage of the ability to pause scenarios as needed to allow for discussion of what was going on with the patient. We expected the students to demonstrate competence in clinical skills that was appropriate for medical students who will be entering the clerkships. Similar to that in our paper-based PBL sessions, the students recorded the information they acquired by asking the patient questions and conducting a physical examination (data), they hypothesized causes of the problem (hypotheses), they explained reasons for their choices of hypotheses and decided what information they needed to help them prioritize their hypotheses (analysis), and the students determined what other concepts or areas of knowledge they needed in order to explain a hypothesis or some other aspect of the case (learning issues). The data, hypotheses, analysis and learning issues were recorded on flip-charts. After the last session, the students and facilitator summarized the content that was learned and reflected on the process.

What the Students Thought About the Sessions

The following comment by a student captured and summarized the feelings of other students after their experience with the simulator not only in terms of their commitment to learn but also to have an opportunity to apply what they have learned in their basic science courses to a clinical situation:

> I have been waiting to learn from Stan [the patient simulator] because I have heard great things from the other students. It was a real challenge to start putting what we had learned in ICM [Intro-

duction to Clinical Medicine, a fourth semester course] and in all other semesters together.

The students had "been waiting to learn from" the simulator or were anticipating the experience. This represents two of the assumptions for adult learning, readiness to learn and orientation to learning (Knowles, Holton, & Swanson, 1998). The timing of our simulator sessions in the fourth semester just prior to their first experiences in their clinical years coincides with the students' move from their preclinical stage to their clinical stage. The students were prepared to learn what they needed to know and do, so that they would be ready for their clinical rotations. The students also were looking forward to their future roles as clinical clerks so that their orientation to learning was directed toward situations that they may encounter. The fact that the simulator "shows you [the students] what you are going to see in your clinical rotations," motivated the students to learn since the experience will help them deal with their clinical experiences.

A frequent comment made by students was that the experience "ties in everything, puts all of the subjects together" and "when you're in here [simulator session] you have to pull it [information] from somewhere . . . ." The preclinical years of undergraduate medical education are designed to provide students with the knowledge necessary to function effectively in the clinical clerkships. Oftentimes, students have difficulty applying what they have learned when at a patient's bedside because the basic science curriculum fosters in students the notion that learning is memorization of a list of facts with little opportunity to, or help with, integrating the information among the disciplines. The simulator sessions provide the context in which students can practice applying what they have learned to clinical situations. Gordon, Oriol, and Cooper (2004) reported similarly that the simulator provided the students with a cognitive framework for integration and recall rather than a lesson in technical proficiency. The students also felt that they "remember things better because it's hands-on instead of just reading it and learning it that way." Additionally, a significant difference between a novice and an expert is that the experts' knowledge is connected and organized around important concepts. The fact that the expert's knowledge is "conditionalized" helps them to know the contexts in which their knowledge is applicable (How people learn, 2000, p. 9). The simulator provides the context and the students apply their knowledge to solve the problem. We found that students often attempted applying knowledge that was irrelevant to the case, which is to be expected at their level of expertise. It is expected that with continued practice in similar situations, either with simulators or in their clinical rotations, the students' knowledge will become conditionalized and they will be better able to retrieve knowledge that is relevant to the task.

Another significant difference between our typical PBL paper cases and the PBL case using the simulator was the added dimension of stress. This was evidenced in our case by a comment from a student during a debriefing session who stated that at the beginning of the first session, while he was just observing, "I saw people frozen. I saw people sweating bullets." This type of experience during their preclinical education is important in that it is preparing them for experiences that they will have with their first patient. As a result of having had this first experience already, they will be better able to maximize new learning opportunities in the clinical rotations. This notion was captured by a student who felt that "when we get into the hospital, we'll have more insight than we would have if we didn't have this experience." The students were confronted with a talking patient who wanted to be treated and was frustrated by being asked numerous questions. When comparing the classroom environment to that of the simulator students commented that "on a test you can skip a question and get to the end of the test go back to it. You can't because he's [the simulator] sitting here bugging you to treat him because he can't breathe" or "make a mistake on paper, you get the wrong answer—make a mistake here, the patient dies." These comments suggest that the added stress was not detrimental to their learning. Rather, it represented a new or motivating experience for students and that they had started thinking of the simulator as a patient. This was summed up by a student who said: "The ability to think on your feet as opposed to having time to sit and process, we have not been in this position before."

The simulator also provided opportunities for our students to develop competencies in other areas aside from knowledge including lifelong learning, interpersonal and communication skills, and professionalism as exemplified, respectively, by comments from three students:

> It showed us we learned stuff but yet there is a lot we don't know. Learning what those gaps might be, we could go do the review and refresher for ourselves so we can be better physicians later on.
>
> A lot of different issues we were able to experience, one of them being teamwork.

> One of the biggest things about being a doctor is being able to deal with people. Here we were able to practice respecting the patient and each other.

The Association of American Medical Colleges has proposed guidelines for medical student education through the Medical School Objectives Project (Association of American Medical Colleges, 1998). Similarly, the Accreditation Council for Graduate Medical Education has developed competencies which medical residents should demonstrate proficiency (Accreditation Council for Graduate Medical Education, 2003). Knowledge of science basic to medicine is an area addressed by both of these organizations. Other areas include those emphasized in the comments made by the three students above as well as interviewing and physical examination skills. Investigators have shown that clerkship directors think students are not prepared for their clerkship years in several areas of competency (Windish, Paulman, Goroll, & Bass, 2004). In a survey of 140 clerkship directors, approximately 80% or more of the clerkship directors felt that preclinical students needed high levels of ability in communication skills, professionalism, and interviewing and physical examination before they entered their clerkships. Thirty to 50% of the respondents felt that students were less prepared than they should be (Windish et al., 2004). The fact that our students can see how the simulator sessions helped them in areas aside from knowledge, attests to the power of the HPS as a tool for helping students achieve desired competencies.

Finally, the students were unanimous in feeling they appreciated the experience and it was fun, which is a pervasive finding in the literature (Forrest & Taylor, 1998). The students also insisted that the simulator sessions should be mandatory for all students and there should be more of them. This is consistent with the findings reported by Gordon et al. (2004).

Is Learning Enhanced?

We have only just begun our simulator program at RUSM and, based on comments from students in our debriefing sessions, there is the perception that learning is enhanced. There is a paucity of studies addressing the issue of whether learning is enhanced. However, many researchers are beginning to examine the role of simulators in enhancing student learning in the preclinical curriculum. Euliano (2000, 2001) at the University of Florida College of Medicine in Gainesville, described the implementation of the simulator, students' opinions on its use, and student performance on topics related to cardiovascular physiology. In an effort to reintroduce laboratory experiences in physiology, which were traditionally done using animals, the faculty developed simulator scenarios that illustrated important concepts in physiology. The workshops were held at the same time the students were receiving lectures on the material such that the workshops with the simulator reemphasized important concepts taught in lectures. In the first paper (Euliano, 2000), the investigator divided the students into groups of 15. During the two-hour workshop, groups of three students worked through one of five cases while their colleagues observed. In the second paper (Euliano, 2001), the students were divided into groups of 10 to work with the simulator in one-hour sessions. In the initial study (Euliano, 2000), the investigator found that students nearly unanimously rated the workshops as "Very Good or Excellent", which was consistent with our findings, and that the students' self-assessment of their understanding of a topic increased after the workshop compared to their understanding of the same topic before the workshop. In a subsequent study (Euliano, 2001), the investigator administered a test before and after the workshop and showed performance improved in five out of the seven questions on the tests.

The National University Hospital in Singapore also has used simulators in teaching concepts of physiology (Tan et al., 2002). The simulations were conducted after the students had completed the blocks on cardiovascular and respiratory physiology. There were 19-25 students in each group. The workshop sessions were 60 minutes long divided into 20-minute-blocks. The first block was a 20-minute didactic lecture followed by a true-false pretest. The next block was a 20-minute session with the simulator followed by a post-test. The investigators of this study demonstrated a 27% increase in test scores after the simulator experience. In terms of students' opinions of the simulator, there was near unanimous agreement that the simulator session was a better way to learn physiology and they would like to learn other subjects using the simulator.

Together the results of these studies and our experiences support the notion that learning is enhanced when students are able to work with a human patient simulator during their preclinical years. Studies conducted on the use of simulators in the clinical setting do provide evidence that acquisition and retention of knowledge are improved by simulation and there is transfer of skills to the clinic (Issenberg, McGaghie, Gordon, Symes, Petrusa, Hart, & Harden, 2002; Forrest & Taylor, 1998; and Issenberg et al, 1999).

Challenges to Implementation and Potential Solutions

Aside from the fact that the educational efficacy of the simulators in learning remains unproved, there are other challenges to implementation including cost, faculty time, and limited student group size. One of the most significant challenges to implementation is the cost of purchasing and maintaining the simulator. The human patient simulator and patient monitor we purchased cost approximately $200,000 and $18,000, respectively (2003 US dollars). There are now basic simulator units available for less than $50,000. Annual expenses of a maintenance contract, gasses, and other supplies will cost approximately $15,000. We also employ a technician who is responsible for maintaining the supplies in the simulator suite and preparing the simulator for each session. Finally, the two authors of this chapter devote a considerable amount of their time to the simulator program which also must be considered. These financial burdens can be defrayed when other departments in the institution share the costs. Other methods to defray costs include obtaining educational grants, offering continuing education programs to health care professionals on a fee basis, soliciting donations, and by charging student laboratory fees.

Faculty contact time with students may increase as a result of the need to facilitate the multiple simulator sessions. In our case, we incorporated the simulator into an existing PBL program which did not have a significant effect on faculty workloads. Faculty members that were facilitating a paper-based PBL case, exchanged their time to facilitate a simulator case instead. We have a unique situation at RUSM in that we have approximately 1,100 students enrolled in the four semesters on Dominica which means the possibility exists that we could be limited by the number of faculty members on staff. This barrier can be minimized with sufficient numbers of faculty and flexible scheduling. Additionally, faculty members can be recruited from hospitals or other health care centers and trained to be facilitators. The examples provided herein demonstrate that the simulator oftentimes is used to reinforce classroom learning. The simulator session could replace the didactic lectures which would mean the students would have to come prepared to work with the simulator. Faculty time also is necessary for case development but cases do not have to be developed from nothing. Clinical cases can be adapted to the simulator from books and web sites. Monetary incentives represent another way to encourage faculty members to develop cases.

Based on our experience, group sizes should not be larger than 10 students so each student has a chance to interact physically with the simulator. If class sizes are large, as in our case, then the students may only get the opportunity to use the simulator one time during the course of a semester. These challenges can be overcome by acquiring additional simulators. More students could be involved in the sessions if they are presented either live or recorded on video monitors in a large auditorium. An alternative approach would be to allow students to work on their own time unsupervised by a facilitator. In this setting, the simulator session would have to be a well-designed laboratory in which the students are provided a step-by-step guide. The results that students generate, when they follow each step in the guide could be used by them to prepare laboratory reports. These last two approaches would diminish some of the benefits of the simulator in terms of student competencies that come from working in a group with a facilitator as discussed above.

Conclusion

In this chapter we described our experiences and students' responses to our fledgling simulator program at RUSM. We offer five take home points for institutions that may be considering instituting a simulator program. First it is helpful to have the support of the dean and president as well as to have champions of simulator use. These individuals probably already exist on all campuses, and they are the early adopters of technology. Second, start slowly. Plan to take your time to get to know the simulator and try it out on a few groups of students. The simulator program can be implemented rapidly if it evolves from an existing program such as a PBL program. Share the results of your trials at every opportunity including general faculty meetings and curriculum committee meetings. Third, cultivate relationships with others across departments. Not only will the expenses be shared over multiple departments but more faculty members will become enthused and want to participate. Fourth, if facilitators are required, make sure that they are trained. There are several universities that offer training in PBL facilitation and workshops are often available as part of national and international meetings on simulation which are held annually. Finally, reinvest the feedback that you get from students back into the program. Our students found this a valuable experience and they provided thoughtful feedback in a desire to see the program flourish.

Many articles published regarding the use of simulators in training physicians often conclude with statements questioning whether simulator use in education leads to better learning. There is, in general, a call to institutions of higher learning to provide data supporting the notion that students learn better with a simulator. In our case, we do not have evidence, as yet, of enhanced learning but it is instructive to point out that students would choose to learn this way over conventional methods of instruction such as lectures. Thus it is difficult to argue that the enjoyment and motivation that the students exhibit during the simulator encounter are not contributing to their learning. We recognize more study on the pedagogical value of simulators is needed but as Gaba (1992) has pointed out, "no industry in which human lives depend on the skilled performance of responsible operators has waited for unequivocal proof of the benefit of simulation before embracing it" (p. 494).

References

Accreditation Council for Graduate Medical Education. (2003). *Outcome project: Enhancing residency education through outcomes assessment.* Retrieved April 28, 2004, from http://www.acgme.org/outcome/comp/compFull.asp.

Association of American Medical Colleges. (1998). *Learning objectives for medical student education: Guidelines for medical schools. January 1998. Report I of the Medical School Objectives Project.* Retrieved April 28, 2004, from http://www.aamc.org/meded/msop/.

Committee on Developments in the Science of Learning and Committee on Learning Research and Educational Practice. (2000). *How people learn: Brain, mind, experience, and school.* Washington, DC: National Academy Press.

Euliano, T. Y. (2000). Teaching respiratory physiology: Clinical correlation with a human patient simulator. [Electronic version] *Journal of Clinical Monitoring and Computing, 16*, 465-470.

Euliano, T. Y. (2001). Small group teaching: Clinical correlation with a human patient simulator. [Electronic version] *Advances in Physiology Education, 25*, 36-43.

Forrest, F. & Taylor, M. (1998). High level simulators in medical education. *Hospital Medicine, 59*, 653-655.

Gaba, D. M. (1992). Improving anesthesiologist's performance by simulating reality. *Anesthesiology, 76*, 491-494.

Gordon, J. A., Oriol, N. E., & Cooper, J. B. (2004). Bringing good teaching cases "to life": A simulator-based medical education service. *Academic Medicine, 79*, 23-27.

Issenberg, S. B., McGaghie, W. C., Gordon, D. L., Symes, S., Petrusa, E. R., Hart, I. R., Harden, R. M. (2002). Effectiveness of a cardiology review course for internal medicine residents using simulation technology and deliberate practice. *Teaching and Learning in Medicine, 14*, 223-228.

Issenberg, S. B., McGaghie, W. C., Hart, I. R., Mayer, J. W., Felner, J. M., Petrusa, E. R., et al. (1999). Simulation technology for health care professional skills training and assessment. *Journal of the American Medical Association, 282*, 861-866.

Knowles, M. S., Holton, E. F. III., & Swanson R. A. (1998). *The adult learner: The definitive classic in adult education and human resource development.* Woburn MA: Butterworth-Heinemann.

McIndoe, A. & Jones, A. (2003). *Worldwide simulation sites*. Retrieved April 28, 2004, from Links and Statistics at the Bristol Medical Simulation Centre web site: http://www.bris.ac.uk/Depts/BMSC/

Milbank Memorial Fund. (2000). *The education of medical students: Ten stories of curriculum change.* New York: Author.

Morgan, P. J. & Cleave-Hogg, D. (2002). A worldwide survey of the use of simulation in anesthesia. *Canadian Journal of Anesthesia, 49*, 659-662.

Schwartz, P., Mennin, S., & Webb, G. (Eds.). (2001). Introduction. In *Problem-based learning: case studies, experience, and practice.* London: Kogan Page Limited.

Tan, G. M., Ti, L. K., Suresh, S., Ho, B. S., & Lee, T. L. (2002). Teaching first-year medical students physiology: Does the human patient simulator allow for more effective teaching? [Electronic version] *Singapore Medical Journal*, *43*, 238-242.

Windish, D. M., Paulman, P. M., Goroll, A. H., & Bass, E. B. (2004). Do clerkship directors think medical students are prepared for the clerkship years? *Academic Medicine*, *79*, 56-61.

Interactive Excel Tutorials in a Quantitative Literacy Course for Humanities Students

Vera Frith
Jacob J. Jaftha
Robert N. Prince
University of Cape Town

Abstract

Inadequate mathematical literacy acts as a barrier to students wishing to study a variety of courses at the university level. One of the strategies we have used to address the lack of mathematical and quantitative reasoning skills in students, is the integration of interactive computer tutorials into the curriculum, created using Excel workbooks coded with Visual Basic for Applications (VBA). The incorporation of a custom-designed graphics tool (programmed using the Excel macro language, VBA) that can be included within an Excel workbook wherever it is needed, has enhanced the design of the tutorials that deal with pre-calculus concepts. A system for recording student responses to a database has allowed for some of the assessment to be automated. The results of observations, course evaluation questionnaires and student interviews indicate that in general the use of these interactive computer-based materials is well-received by our students.

We review some of the literature dealing with the advantages of computers to enhance the teaching and learning of mathematics. We briefly describe the results of a systematic study of the effect of the course, and the computer component in particular, on our students' attitudes towards and feelings about learning mathematics and using computers. We also discuss the results of another study we conducted into the effect of the timing of the delivery of the computer tutorials (relative to the classroom materials), making use of pre- and post-testing of students' understanding of concepts.

Introduction

At our university, it is recognised that many students have inadequate quantitative literacy and mathematical skills to enable them to cope with their chosen course of study (Brink, 2001). Apart from the fact that the traditional approach to teaching mathematics in schools does not develop sufficient levels of quantitative literacy (as discussed by Hughes-Hallett, 2001), in South Africa there is a legacy of educational disadvantage that still affects the majority of the population. Under Apartheid, there was an explicit policy of denying black students access to mathematical and scientific knowledge, and it will take many years to reverse the effects that this has had on the education system.

For disadvantaged students who are studying economics-related and other quantitative subjects in the Humanities Faculty, the Mathematics Department provides a first-year course entitled "Effective Numeracy." This course has the objective of increasing students' levels of quantitative literacy, supporting their studies in the rest of their programme and preparing them for mathematics and statistics courses in later years. Brink (2001) and Frith and Prince (2001) have outlined the philosophy of the course and its development over a period of time. One of the most important principles in the design of the course is to create a non-threatening co-operative environment for learning that will allow students to develop confidence in their ability to succeed, as many of these students have a high degree of "maths-anxiety." The very wide range of ability and prior experience, particularly in the use of computers, within the classes, poses a significant challenge to the realisation of this objective.

Currently the course consists of two semesters with slightly different objectives. In the first semester the emphasis is on developing quantitative literacy through the use of context-based applications. The second semester of the course content is closer to a more traditional pre-calculus (bridging) course. One third of this course throughout the year (in terms of classroom time and credits) is conducted in the computer laboratory through the medium of Excel-based interactive tutorials that are tightly integrated into the curriculum. In the first semester there is an emphasis on learning to use Excel in "real-life" contexts to perform data analysis and to represent quantitative information graphically. In the second semester the Excel tutorials are intended to support the learning of mathematical concepts, in particular, the idea of functions, graphs and gradients.

Excel Based Interactive Tutorials

In designing the Effective Numeracy course, we have assumed that using computer-based tutorials will enhance students' understanding of the quantitative concepts. There are numerous reasons to believe that properly used computer tutorials can add value to a mathematics course, and these have been comprehensively reviewed by Kaput (1993). In particular, it is the interactivity of the computer as a medium for learning that distinguishes it from other available media. Kaput (1993) describes two general types of interactions that such systems can provide: a) built-in constraints or supports and b) user-directed agents that perform actions for the user. Examples of the first category are the automatic choosing of a suitable scale by a graphing package, or the overt feedback or commands programmed into educational software. Examples of user-directed agents are the facilities for performing computations, drawing graphs, recording results, sending messages, and searching and displaying required information.

The most obvious advantages that are exploited in our course are the computational and graphical abilities of Excel, which allow more examples to be done by the student and also permit the use of more realistic values in these examples. The Excel environment is ideal for learning about data analysis and representation using real-life examples, which is strongly emphasised in the first semester quantitative literacy curriculum. In the bridging component, the use of Excel facilitates the understanding and representation of functions in the four different ways that the course emphasises: with a formula, with a table of values, graphically and verbally (The "Rule of Four"of Hughes-Hallett, Gleason, McCallum, et al., 1998).

The computer also makes it possible to illustrate certain concepts and processes graphically in a dynamic manner, which is difficult to achieve on paper or a blackboard, and which we believe assists the students to develop the ability to produce their own mental images in situations where they are helpful. Obviously we also believe that there is an advantage in approximating the "conversation" of an individual tutorial situation where each student receives immediate feedback as they work through a computer tutorial at their own pace.

There is a great deal of experience of and knowledge about using Spreadsheets to enhance learning in mathematics, some of which can be accessed through a website at Vienna University (Neuwirth, 2001). Some practitioners have also used extensive Visual Basic for Applications (VBA) code to create Excel-based tools and environments that can be used to enhance learning (for example Carr, 2000). Our approach is to make use of VBA code within an Excel workbook, to program self-contained "tutorial-simulations" (Laurillard, 1993) that can be used semi-independently by the students without the support of any other materials.

A typical tutorial will consist of several electronic Excel worksheets, one stating objectives, some containing the interactive presentation of relevant mathematics content, and others comprising examples and exercises, many of which make use of a custom-built "graphics tool." This allows the student to produce the graph of a function (as they would with a separate graphics package), but at the relevant place within the worksheet and without interrupting the interactive conversation that comprises the tutorial.

Some of the advantages of using Excel as an authoring environment are that the students are familiar with it, that all the functionality of Excel remains accessible to the students throughout the tutorial, and that Excel is so commonly used that the tutorials become extremely portable. In addition it is very easy in this environment to change the content, for instance to modify an example or introduce different data. It is also relatively easy to program fairly sophisticated interactions and animations using the in-built VBA macro language, and to record students' answers to a database for automated processing.

Integration of Tutorials into the Course

Every week the students spend one 2-hour session in the computer laboratory, and two 2-hour periods in the workshop-lecture environment. The actual mathematical material dealt with in the computer tutorial for any particular week is always covered in the classroom within that same week, and the lecturers concerned are encouraged to make the links between the laboratory and the classroom materials explicit.

There is a critical relationship between assessment practices and the nature of student learning (Luckett and Sutherland, 2000) which means that planning the assessment structure and practices is an integral part of the curriculum design for computer interventions as well. The continuous assessment process built into the design of the entire course (which has both a formative and a summative purpose) is also applied to the computer component. To assist with this process, a system has been developed whereby students' responses to questions in the tutorials and assessment can be recorded automatically to a database. This data is then processed to produce feedback to individual students about their misconceptions, and to the tutors about the class's performance in general. This system also can be used to automate some of the burden of marking that the regular evaluation schedule generates.

Every third week there is a "computer evaluation" in which students complete a computer tutorial that is submitted for marking and counts towards their class record. As pointed out by Laurillard (1993, p. 217), use of computer learning material "must be integrated with other methods, the teacher must build on the work done and follow it through, and most important, the work students do on the material must be assessed." One third of the final examination for this course is also conducted in the computer laboratory through the medium of an Excel workbook similar to the tutorials, but without feedback.

The ability of the computer-based tutorial to provide the student with immediate feedback (and a "conversational" tutorial environment), has already been stressed as one of the main motivations for using this type of intervention. In the Excel tutorials, in almost every case where a student provides an answer to a question, they will immediately receive explanatory feedback. Ideally this feedback would be "adaptive," allowing the tasks presented to the student to be tailored to their particular needs, as manifested by their performance on questions in the tutorial (Laurillard, 1993). This level of interactivity is not built into our Excel workbooks, but is made available through the presence of the lecturers and tutors in the laboratory sessions. All students perform the same tasks on the computer, (although they can control the order), and more advanced optional tasks are also made available. Students who need additional support can get almost immediate assistance from a tutor, of whom there are 3 present in each class of less than 30 students. It is not our intention that these tutorials be used for independent study, but rather to support the learning of material that is also dealt with in workshop-lectures.

As mentioned before, the Effective Numeracy course concentrates on quantitative literacy in the first semester and on pre-calculus in the second. During this semester, students explore the properties of different functions and use the graphical capabilities of Excel to solve "word-problems" graphically. Rather than draw up a table of values for every function they wish to represent, or change to another environment such as a graphics calculator, they are provided with a pre-programmed "graphics tool." which allows them to create up to five graphs simultaneously merely by typing in the formulae. A separate instance of this tool is placed at the appropriate place in the tutorial, wherever it is needed, so that producing a graph can be done in a manner that creates the least distraction to the student's train of thought while working through the tutorial.

A further advantage of the "in-situ" graphics tool is that when marking a student's work, the lecturer can see the graphs that the student actually plotted. This allows for partial credit being given for partly correct work, and allows the lecturer to develop an understanding of students' difficulties and misconceptions. If students were using a graphics package or calculator "on the side," one would only have access to their final answers for evaluation purposes.

Students' Response to the Tutorials

Do the students believe the tutorials helped them to understand the mathematics content of the course?

Every year since 1999, the students in the Effective Numeracy have completed a comprehensive course evaluation questionnaire at the end of the first and the second semester of the course. In addition a randomly selected sample of students were interviewed about their reactions to the course at the end of 2001.

In response to the questions on forms filled in by about three quarters of the class at the end of the 2001 course (N = 74), 70% of the respondents claimed to have found the explanations contained in the tutorials useful, while 82% were positive about the usefulness of the feedback in improving their understanding. One aspect of the tutorials that many students "particularly liked" was the automated drawing of graphs. Others mentioned the feedback, the visualisation of concepts, and the fact that they were "doing maths" on computers.

In 2001, extensive interviews were carried out with 11 students, chosen at random, in which one of the questions asked was: "Did the laboratory tutorials contribute to your understanding of the maths?" These were some representative responses:

- "Yes, definitely. The class and labs helped each other. For example the labs helped with understanding graphing and the class dealt with the equations, which were needed in order to do the labs and so on. Both helped each other."
- "Yes, if you don't understand in class, it comes together in the labs. The feedback is very useful. It shows how the lecturer would answer the same questions. . . . (The labs) really brought about a greater understanding. What we did in class we would get immediate practice in the labs."
- "Labs were very visual. You can actually see how things work. In class you have to imagine for yourself what things look like–in the labs you can see it clearly before your very eyes, it is easier to understand. They are helpful, they show more than you see in class–especially the graphics package (I explored more than the tut told me to)"
- "Ja, it did, but mostly to computer skills. . . . Labs gave direction for solving word sums, helped for doing similar ones again later."
- "The good thing about the labs is that even if you don't want to ask questions, you still get feedback. There are lots of exercises and real-life examples. They helped you through the problems and helped you to understand how to integrate concepts."
- "Yes, they provide reinforcement to what was done in class, step-by-step. I liked learning how to use the computer."

Seven of the eleven students felt strongly that the computer tutorials had helped them to understand the mathematics content of the course, and there were also seven who remarked on the interdependence of the laboratory and classroom materials. However there was a tendency to see the laboratories as reinforcing work done in the classroom, rather than the other way around. This is consistent with the observation that students prefer to see new ideas first in the classroom. Two of the students specifically referred to the feedback as being helpful with learning the mathematics and two remarked on the usefulness of the step-by-step approach utilised in the tutorials for solving "word sums." It is interesting that one of the students saw the visual (graphical) nature of the tutorials as making the greatest contribution to her understanding.

Do the tutorials have any effect on students' attitudes to learning mathematics (and to using computers)?

This is a difficult question to answer because of the close integration of the tutorials into the course. Any recorded attitude changes cannot be ascribed to any particular component of the course. However there are indications from the course evaluation results, that there was a general improvement in students' feelings of confidence in doing mathematics (and in using computers).

To investigate attitude changes more thoroughly, a study was conducted using the properly validated scales described by Cretchley et al (2000). The scales developed by these researchers were used to measure the confidence (both with mathematics and with using computers) of students in the Effective Numeracy course, at the beginning and at the end of the first semester. Some relevant findings were (Frith, Jaftha and Prince, 2004a):

- there is in general a very low correlation between confidence in doing mathematics and confidence with using computers
- over the duration of one semester there is no significant change in the confidence (of the class considered as a whole) measured by these scales.
- the female students were significantly less confident with using computers than the male students
- there is a moderate correlation between the change in mathematics confidence and the change in computer confidence over the semester.

Does the timing of the delivery of a computer tutorial relative to the related classroom sessions have an effect on the learning?

Since there are three lecture groups who attend the laboratories on different days of the week there is variation in the experience of the students in different groups as far as the timing of the computer tutorials is concerned. Some students will encounter a new idea for the first time in the laboratory and then have it reinforced in the classroom, while others see the same idea for the first time in the classroom, (and then have it reinforced in the laboratory.

Our observations and students' comments indicate that some students in our course are more comfortable when they encounter new knowledge on paper first, and then apply that knowledge in the computer tutorials. It is possible that it is easier to assimilate new ideas (and transfer them to other contexts) when they are first introduced in a medium that is familiar.

We conducted a study to gain insight into whether the timing of delivery of the computer tutorials has an effect on learning, by performing pre- and post-tests on concepts in the case where the concept was first introduced in the laboratory and in the case where it was first encountered in the classroom. The intention was to explore whether it is possible to observe differences in the nature of the learning between students in the classes who experienced the different learning environments in a different order.

Initially the results were inconclusive, but provided a great deal of insight into the requirements for the design of such an experiment, with the most noteworthy result being that the medium in which the pre-test was delivered had a very significant effect on students' performance. After several iterations of the experiment on different tutorials, during which we refined our techniques, the data did reveal real differences in students' learning experiences in the classroom and in the computer laboratories. It appeared that in some respects the computer tutorials were more effective in conveying the concepts than the classrooms were. The results of this investigation are published in detail elsewhere. (Frith, Jaftha and Prince, 2004b)

These studies also reinforced our belief that language difficulties play a major role in contributing to the problems experienced by disadvantaged students in learning quantitative subjects. It is impossible to separate the understanding of the quantitative concepts from the understanding of the terms used to define them and the competent use of language to express that understanding. Language, especially in the South African context, is always going to be a complicating factor in the learning of concepts in a foundation course such as the Effective Numeracy course, and plays a major role in the state of "disadvantage" experienced by many of our students, a situation which we believe the integration of interactive computer tutorials into the course helps to alleviate.

Summary

The "Effective Numeracy" course at a South African University aims to make studying mathematics an enabling experience that provides students with the tools and competencies that they really need to succeed in their course of study. The integration of interactive computer tutorials into the curriculum is one of the major strategies we have used to achieve this.

The results of observations, course evaluation questionnaires and student interviews indicate that in general the use of interactive Excel-based tutorials for supporting the learning of mathematics concepts is well-received by the students in the Effective Numeracy course. They appear to have a positive opinion of the value of the computer tutorials in contributing to their learning.

The incorporation of a custom-designed graphics tool (programmed using the Excel macro language, VBA) that can be included within an Excel workbook wherever it is needed has enhanced the design of the tutorials used in the pre-calculus part of the course. A system for recording student responses to a database has allowed for some of the assessment to be automated.

As the ability to learn mathematics is strongly influenced by affective factors, especially confidence and anxiety, we explicitly try to address these factors in our curriculum design. A systematic study of the effect of the course, and the computer component in particular, on students' attitudes towards and feelings about learning mathematics and using computers was carried out. The quantitative instrument used to survey students' attitudes was not able to reveal significant changes in attitude for the class as a whole over one short semester, but did allow those students whose confidence and attitudes had changed significantly to be identified.

A study of the effect of the timing of the delivery of the computer tutorials (relative to the classroom materials), making use of pre- and post-testing of students' understanding of concepts, showed that it is possible to reveal differences in the learning experience in the different settings. These observations provide insights that will inform the design of computer tutorials that more effectively meet the needs of students in the Effective Numeracy and other courses in the future.

References

Brink, C., (2001). Effective numeracy. *Transactions of the Royal Society of South Africa*, *54*(2), 247-256.

Carr, R., (2000). *XLMathematics*. Retrieved June 19, 2003, from http://www.deakin.edu.au/~rodneyc/XLMATH.HTM

Cretchley, P., Harman, C., Ellerton, N., & Fogarty, G., (2000). MATLAB in Early Undergraduate Mathematics: An Investigation into the Effects of Scientific Software on Learning. *Mathematics Education Research Journal*, *12*(3), 219-233.

Frith, V., Jaftha, J. J. & Prince, R.N. (2004a). *Proceedings of the 12th Annual Conference of the Southern African Association for Research in Mathematics, Science and Technology Education.* Durban: SAARMSTE.

Frith, V., Jaftha, J.J. & Prince, R.N., (2004b). Evaluating the effectiveness of interactive computer tutorials for an undergraduate mathematical literacy course. *British Journal of Educational Technology*, *35*(2), 159-171.

Frith, V., & Prince, R.N., (2001). Gatekeeper vs. Gateway, *Communications of the Third Southern Hemisphere Symposium on Undergraduate Mathematics Teaching*, 46-50.

Hughes-Hallett, D., Gleason, A.M., McCallum, W.G., et al. (The Consortium based at Harvard), (1998). *Calculus. Single and Multivariable* (2nd ed.), New York: Wiley.

Hughes-Hallett, D., (2001). Achieving Numeracy: The Challenge of Implementation. In L. A. Steen (ed.), *Mathematics and Democracy, The Case for Quantitative Literacy* (pp.93-98). USA: The National Council on Education and the Disciplines.

Kaput, J. J., (1992). Technology and Mathematics Education. In D.A. Grouws (ed.), *Handbook of Research on Mathematics Teaching and Learning, A project of the National Council of Teachers of Mathematics* (pp. 515-556). New York: Macmillan.

Laurillard, D., (1993). *Rethinking University Teaching: A Framework for the Effective Use of Educational Technology.* London: Routledge.

Luckett, K., & Sutherland, L., (2000). Assessment Practices that Improve Teaching and Learning. In *Improving Teaching and Learning in Higher Education. A Handbook for Southern Africa* (pp. 98-130). Johannesburg: Witwatersrand University Press.

Neuwirth, E., 2001, "Spreadsheets, Mathematics, Science and Statistics Education, Quite a Lot of what you always wanted to know." Retrieved June 19, 2003 from University of Vienna website: http://sunsite.univie.ac.at/Spreadsite/

The Virtual Student: Using Technology to Teach School Psychology

Ari S. Yares
T. Chris Riley-Tillman
Catherine A. Fiorello
Marie C. McGrath
Temple University

Abstract

Diagnosis of academic and behavior problems is a key role of school psychologists. Extensive training, including classroom experience and practica, are required prior to unsupervised practice. Because of the ethical dilemmas inherent in allowing incompletely trained school psychology graduate students to practice diagnosis on actual children with learning difficulties, a basic simulated approach using e-mail and case studies is reviewed which would allow for enough practice to gain fluency. Student reaction to this simulation was positive, but it did not allow for adequate rehearsals for fluency to be gained. A more advanced, web-based diagnosis simulator using a database of available cases is discussed.

The Virtual Student: Using Technology to Teach School Psychology

According to Fagan and Wise (2000), "a school psychologist is a professional psychological practitioner whose general purpose is to bring a psychological perspective to bear on the problems of educators and the clients educators serve" (p. 4). Most school psychologists are employed by public schools and guide students through the eligibility process for the special education system. Typically, a school psychologist spends a significant amount of time completing psychoeducational assessments of students with academic or behavioral problems to determine eligibility (Fagan & Wise, 2000). The school psychologist assesses a student's academic skills, learning aptitudes, personality and emotional development, and social skills, and observes the learning environment and school climate. Through an appropriate psychoeducational evaluation, a school psychologist can determine which barriers are interfering with a child's education and suggest ways of overcoming them, either through regular education or through special education.

The vast majority of the literature on diagnostic decision-making stems from the field of medicine. An earlier study used a "systems-analytic" approach to describe the diagnostic process used by psychiatrists (Nathan, 1967). It used a flowchart to parse out the decision-making process used by psychiatrists as hypothesized by Nathan (1967). The system has been developed based upon current practice in psychiatry at the time of the study and was assumed to be the preferred model for diagnosis.

Nathan's (1967) results indicated that the most critical stage for differentiating between diagnoses lies at the beginning of the diagnostic process. If, as Nathan (1967) proposed, more experienced psychiatrists would function more closely in line with the model, they may be more aware of the potential diagnoses and differentiate correctly more often than less experienced diagnosticians. Likewise, they would move through the diagnostic process in a more efficient manner as they eliminated options and reduced the need for unnecessary assessment tools.

A variety of recent studies have evaluated the diagnostic process using simulations. Voytovich, Rippey, and Jue (1986) note that many of these simulations are problematic as they have been developed around theoretical patients with a single diagnosis. Patients such as these are not the rule in medical practice as patients often present multiple symptoms that do not point to a specific diagnosis.

Papa, Aldrich, and Schumaker (1999) have addressed this concern by exploring differential diagnosis. They note that differential diagnosis performance requires experience with an extensive number and variety of cases. Differential diagnosis is also very dependent upon the knowledge of specific diseases and problems.

The majority of research on diagnosis in school psychology has focused on issues other than the diagnostic process. These issues range from compliance with federal laws and other mandated regulations to the impact of biases on diagnostic decisions (Johnson, 1990). Johnson's (1990) analysis of several studies of the diagnostic process in school psychologists confirmed Nathan's results (1990). She found that in each study the more experienced diagnosticians needed less time to accurately complete their diagnosis than their colleagues with less experience. Other variables, such as the amount of information requested and confidence in diagnostic decisions did not consistently appear to differentiate between groups of school psychologists.

The skills required for diagnosis within a school setting are extensive. According to the National Association of School Psychologists (NASP, 2002), at least two years of full time graduate level education and a year long internship are required to become a masters level school psychologist. Currently, most school psychology students are trained through a series of lecture-based classes and practicum experiences. Lecture-based courses usually include basic assessment techniques, behavior modification, and education theory. Practicum experiences give students an opportunity to apply knowledge gained in lecture-based courses. At this point, it is assumed that the students have mastered the requisite skills and complete a year long, school-based internship that gives the students the opportunity to practice their skills in novel situations.

For example, in Temple University's School Psychology Program, training in diagnosis occurs in two settings. The first, 'Introduction to the Diagnostic Process,' is a lecture-based course taught during the first year of training. In the course, students are introduced to the framework that will guide their assessments. Hypothesis generation and the diagnostic criteria for various disorders and state eligibility requirements are taught. The next stage occurs during students' second or third year in the program when they work as clinicians in the university-run psychoeducational clinic. Faculty members who are certified and licensed as school psychologists by the appropriate state agencies supervise the trainees. In this setting, the student clinician receives a referral that describes an educational or behavioral problem experienced by a client. With the guidance of the supervisors, the clinician determines appropriate assessment techniques and conducts the evaluation. At the end of the process, the clinician meets with the client's parent at a feedback meeting and provides a psychoeducational report.

The Temple psychoeducational clinic requires intense supervision. Only ten trainees can participate in the clinic at a time and they require supervision because, legally, the trainees may not practice without direct supervision of a certified school psychologist. Each trainee is monitored by closed circuit television and videotaped. Each three-hour session with a client is followed by three hours of supervision where the trainees meet as a group with the faculty to discuss cases. These safeguards and the fact that the trainees are still learning the process of diagnosis create a system where each student handles no more than four cases per year.

Another contributing factor to the low caseload carried by clinic students is the amount of time that a novice clinician needs to process a case. In addition to the six hours of class time dedicated to the clinic experience, trainees spend several more hours each week scoring answer sheets for the various tests from the previous session, observing the client at their school, and writing up their findings, observations, and hypotheses in reports. Beyond this time, each student spends several more hours planning the next testing session, which is a week away. Compare this to the average practicing school psychologist who spends approximately 17 hours per case (Fischetti, 2000) which includes testing, interviewing, and writing the report. While some of the transition to the much higher caseload must occur during internship, preparation during initial training is important.

Thomas, O'Connor, Albert, Boutain, and Brandt (2001) note that practicum experiences attempt to provide students with an appropriate range of situations. Because of the real world nature of a practicum, however, they hold that no matter how well structured the practicum experience is, it is impossible to guarantee that each student will have all of the appropriate experiences. In other words, a trainee may be exposed to a client with a learning disability and a client with mental retardation, but may not see a client with attention deficit/hyperactivity disorder (AD/HD).

Are four cases a year enough training for a graduate student in school psychology to receive before starting an internship, and working as a school psychologist in a less intensively supervised fashion? To gain an advanced skill, it is necessary to have a certain amount of deliberate practice (Issenberg et al., 1999). Deliberate practice is the opportunity to perform a specific task at an appropriate level of difficulty with feedback and the opportunity to correct errors and repeat the tasks. A clinic setting could provide deliberate practice, but the subjects required for practice are humans and could possibly be harmed by those unprepared for a practicum setting. In comparison to the four cases experienced by school psychology trainees, teachers practice planning instruction a much larger number of times during their three to four month long student teaching experiences.

Many internship supervisors have found advanced students starting their internships unprepared to practice independently. Fagan and Wise (2000) note that the accreditation standards set by the American Psychological Association (APA) and NASP have created intense program curricula and current restrictions imposed by state regulations do not allow additional course work to be added to advanced degree programs. Since changes to state regulations seem unlikely and it would be unethical to drop the stringent demands placed upon school psychology trainees, other solutions are needed to provide school psychology trainees with the appropriate experiences without increasing the course load or changing the model used to educate school psychologists. The purpose of this chapter is to describe solutions to this problem created by students and faculty at Temple University's school psychology program that use technology to enhance opportunities to gain diagnostic fluency.

An Alternative Training Method

A variety of other professions already require simulated experiences prior to encountering live clients or patients. Issenberg et al. (1999) reported that simulation is being used for training military personnel, airline pilots, business executives, and in a variety of medical fields. These fields make use of simulations because of the potential danger created by sending an inadequately prepared practitioner into the field and the potential costs associated with that inexperience. The military would not risk having a pilot land a billion dollar fighter jet on an aircraft carrier unless they were reasonably sure by testing through simulation that the pilot was capable of the feat.

Medical education was confronted by the same issues facing trainers of school psychologists, namely the need for an education tool that allows for opportunities for deliberate practice of skills in a consistent manner where no one comes to harm (Issenberg et al., 1999). The authors stress that simulators can be adjusted to fit curricular needs, are predictable, and allow for a consistent experience for all trainees. Practica seldom have as much flexibility or control over the types of situations that one experiences. Issenberg et al. (1999) note that simulations gave medical students additional deliberate practice, i.e., an opportunity to undertake a well-defined task at an appropriate level of difficulty with opportunities for feedback, repetition, and correction of mistakes. Simulations are also not complicated with issues such as embarrassment, stress, unpredictable behavior, and outside commitments that are inherent when working with real patients.

One of these simulators is "Harvey," the cardiology patient simulator, which is currently being used at over 90 medical schools to teach bedside skills for diagnosing cardiology disorders. Harvey, a mannequin and computer combination, is capable of simulating 27 different cardiac conditions and their physical symptoms. Additionally, Harvey comes with a curriculum of cardiovascular diseases and other support materials to aid in instruction. Uses of Harvey have varied from teaching beginning students how to take blood pressure and recognize a heart murmur to training bedside diagnostic skills of primary-care physicians (Issenberg et al., 1999).

Just as Harvey has created the opportunity for medical students to ethically gain fluency in their skills, the possibility exists for school psychologists to have the same opportunity. A simulation of the diagnostic process for school psychologists would increase the number of potential cases that a trainee experiences without problems created by practica. Trainees would gain additional practice in organizing an assessment battery and planning an evaluation. By seeing more clients, even if they are simulated, trainees would gain increased familiarity with the tools at their disposal in the same fashion that airline pilots become more familiar with the cockpit in their simulations. Also, through simulation, trainees would be exposed to unusual cases and conditions that they would not get the opportunity to experience while in a training program's clinic.

Educating any professional who incorporates diagnosis in their practice, including school psychologists, is a complex process that requires elements of technical skill, a large knowledge base, and techniques for facilitating the integration of science and practice. Tillman, Fiorello, and McGrath (2001) identified three steps in educating school psychology trainees to assess psychological disorders:

1. Knowledge obtained through didactic means

a) Knowledge of psychoeducational tools, including formal and informal assessment, observation, interviewing, and intervention development and monitoring.

b) Knowledge of disorders and the diagnostic process, including diagnosis using the Diagnostic and Statistical Manual of Mental Disorders–Fourth Edition–Text Revision (DSM-IV-TR); educational identification under the Individuals with Disabilities Education Act (IDEA), Section 504 of the Rehabilitation Act, of the Americans with Disabilities Act (ADA), and the appropriate state criteria; and knowledge of a problem-solving model.

2. Fluency gained through repetition and feedback
    a) Fluency (accuracy and speed) in administration of psychoeducational tools in a controlled setting.
    b) Fluency in incorporating data into diagnostic decision-making in a controlled setting.
3. Expertise gained through supervised practice
    a) Clinical expertise gained through conducting the psychoeducational diagnostic process in increasingly independent settings (e.g., practica, internships, on-the-job supervision) with feedback from supervisors.

The system that Tillman et al. (2001) describe reflects the instructional hierarchy model which divides learning into several stages: acquisition, fluency, generalization, and adaptation (Haring, Lovitt, Eaton, & Hansen, 1978). Each of these skill levels has instructional procedures associated with it that foster skill mastery at that level, such as drill for fluency building and training in the natural context for adaptation (Daly & Martens, 1994).

In applying this to the diagnostic training of school psychologists, it is clear that the first, third, and fourth steps have established teaching methods within school psychology. Knowledge acquisition incorporates the first stage of rote learning (Tillman et al., 2001) of the knowledge base considered as necessary to diagnose psychological disorders, e.g. DSM-IV-TR criteria and familiarity with psychoeducational assessment tools. These skills are gained in introductory courses where students study the psychological disorders. Students also gain these skills by practicing administration of various intelligence, achievement, and socio-emotional tests to friends and family for the purposes of learning the rules and processes for administration.

The third step, generalization, is equivalent to expertise gained through supervised practice (Tillman et al., 2001). At this point in a school psychology student's training, they have are working with real people with real problems. This includes basic practicum where students implement academic or behavior interventions with students in a public school. It also includes more advanced practicum such as working in the program's psychoeducational clinic as a clinician. Supervision during these experiences is very intense. In the clinic, it includes closed circuit television monitoring and group supervision.

The move to adaptation, the fourth step of the instructional hierarchy, occurs during the internship. Students are placed with school districts or agency and expected to work as school psychologists. Supervision is provided, but not with the intensity of the clinic. Students conduct evaluations without direct monitoring. Instead, they are expected to make decisions based upon the knowledge and experiences gained during the previous stages of learning.

It is the second step, the gaining of fluency in the diagnostic process, however, where current training models are lacking. This second stage should incorporate drill and reinforcement of the diagnostic process to facilitate gaining fluency, but it is not addressed as part of the training process in most school psychology programs. School psychology students should be given opportunities to repeatedly practice forming hypotheses and planning evaluations in order to gain fluency with these skills.

Tillman et al. (2001) note that a variety of ethical and practical issues prevent school psychology trainees from gaining needed fluency in the diagnostic process. Most practice by school psychology trainees of basic assessment skills, such as intelligence and achievement testing, uses children without disabilities because inexperienced trainees cannot ethically be sent to treat children who may potentially have a disability. An alternative would be a supervised practicum, but these are time intensive on the part of the faculty and would be required in large amounts in order to create enough opportunities for drill and reinforcement to yield fluency.

The Solution

At Temple University's School Psychology program, the faculty has added a technology component to the Introduction to the Diagnostic Process course to address this problem. In order to give each student the opportunity to practice the diagnostic process, the course included an experience with "virtual children." These "virtual children" possessed the characteristics of children who have been referred to a school psychologist. These imaginary children eliminated the ethical and practical considerations that need to be accounted for with real children. The trainees interacted with the "virtual children" by e-mailing the instructor. These "virtual children" consisted of the following information stored in a database:

> name of child; diagnostic category (if any); referral and background information on the child; stock responses to expected parent and teacher interview questions; results of prior interventions and direct observations; cognitive assessment, achievement testing, and socio-emotional behavioral assessment data; and the results of any interventions carried out as part of the assessment process (Tillman, Fiorello, & McGrath, 2001, p. 6).

If a trainee requested additional information not already included in the database, the instructor created it, keeping in mind the final diagnosis of the child. Throughout the course of the assignment, trainees practiced generating hypotheses as they explored the potential diagnoses of their virtual child. As they requested information, they needed to develop rationales for their requests. In the project's wrap-up e-mail, students assigned diagnoses based on both state and DSM-IV-TR criteria (if applicable) as well as the rationale behind their final diagnostic decisions. Students are also required to present their cases and the reasoning behind their diagnoses (or lack thereof) at individual case presentations. In these case presentations, the student's work with the case is discussed, and suggestions for future improvement are made.

Student Reactions

Turning to technology to assist in solving this dilemma seemed obvious. Technology has already being used in a variety of different ways in psychology education. Professors have used web sites for posting notes and assignments and also used listservs for disseminating information or for promoting discussions outside of the classroom. Recently, the National Association of School Psychologists has created several on-line continuing education courses for students and practicing psychologists.

Because this project used a technology that students were comfortable with, i.e., e-mail, it was readily accepted by the students enrolled in the Introduction to Diagnostic Processes course. Tillman et al. (2001) report that during the first two years of the project, trainees found it to be an integral activity in their development as school psychologists and that it has increased trainee confidence before entering Temple's psychoeducational clinic as clinicians. Additionally, they report that clinic supervisors find that clinicians who participated in the virtual child project have a stronger grasp on the diagnostic process than those who did not. Unfortunately, all of this evidence is anecdotal and lacks an empirical basis.

Another benefit of utilizing an interactive database as the source of all clinical data for this project is that it is easily adaptable, facilitating inclusion of currently accepted "best practices" in school psychology. For example, the data in their current form allow for a Cattell-Horn-Carroll theory (Flanagan & Ortiz, 2001) based cross-battery approach to assessment, which complements trainee cognitive assessment training. The database also allows for the incorporation of academic and/or behavioral intervention data into the assessment process, thus permitting trainees to go beyond formal standardized measures when designing comprehensive assessments. Future innovations in assessment, in addition to minor changes due to test revisions, can be easily incorporated into the database with a few quick keystrokes, thus making it a useful tool for years to come.

While highly interactive, the "virtual child" project has its limitations. Because of the need to answer each e-mail from the trainees personally, a trainee can treat only one virtual child during a given semester. This limits the opportunity for fluency to be gained through repetition. As suggested by Daly, Lentz, and Boyer (1996), an intervention is needed that specifically focuses on the fluency process by giving ample opportunities for drill. In addition to the limited number of opportunities for multiple repetitions with a particular diagnosis, there are no opportunities for a number of experiences with a diverse group of diagnostic problems. This is critical in that students need to become fluent with both a particular type of diagnosis, and the more ecologically valid situation where a number of children with diverse problems are experienced. Furthermore, the extended nature of the project does not reflect the much faster pace of a "real world" assessment. For a practicing psychologist, an actual assessment would take place over a series of days and be followed by a written report. The delay created by the need for the instructor to e-mail the students takes much longer. It also eliminates any sense of urgency that a real child's assessment would bring with it.

A More Interactive Solution

Discussions with the faculty members responsible for developing the virtual child project have yielded a desire to make the project more interactive and less dependent on teacher-trainee interaction for the technical aspects of the project. This would allow the faculty to concentrate on critiquing the diagnostic process taken by the trainee, rather than on the mechanics of cutting and pasting assessment results into e-mails. Additionally, we are considering adding a tutorial feature to the project to reinforce the diagnostic criteria and process of differential diagnosis that is emphasized in the class.

To replace the interaction via e-mail, students and faculty at Temple are developing a web-based version of the project that would offer the opportunity for trainees to diagnose multiple cases in a much shorter time frame. The information for each "virtual child" has been moved into an Access database to better organize the data. This database will then interface with a series of web pages designed using ColdFusion.

A trainee using DTS would log into the software by entering a unique username and password. After a successful login, the trainee would be able to either pick a diagnosis training case or enter the tutorial section of the software to learn about one of the diagnoses from the DSM-IV-TR, e.g. mental retardation, attention deficit/hyperactivity disorder, or learning disability. For example, a trainee who selected mental retardation from the list of diagnoses to explore would be able to read about the different aspects of mental retardation, such as diagnostic criteria, prevalence rates, or degrees of severity. After reviewing the information, the student could take a short quiz to reinforce the material. Quizzes that cover multiple diagnoses would also be available.

The next step for the trainee would be to select a practice case to diagnose. An initial referral question would then be presented and the trainee would begin to formulate hypotheses for the diagnosis. At the diagnostic information menu, the trainee would be able to select from a wide variety of information, e.g. teacher interviews, cognitive testing, direct observation, and interventions, to assist in the diagnostic process. As the trainee selects additional information, he or she can refer back to the tutorial section to refresh their memory of the potential diagnoses as they begin to rule out individual hypotheses. Finally, when the trainee believes that he or she has reached a conclusion, the trainee can select a diagnosis from the diagnosis menu.

Feedback would be available in two ways. First, each information component will have a simulated time value attached to it. For example, conducting a series of behavioral observations would require 1 hour of simulated time while a neurological evaluation would require two weeks of simulated time. The website will use the Access database to track both real and simulated time to encourage users to complete the diagnostic process in the most efficient and effective manner. Students will be able to see how much simulated time has elapsed while they were conducting their evaluation. At the end of the process, the database will report total simulated and real time elapsed.

Similarly, the system will also track the diagnostic decisions entered at the end of each session. After entering their determination, the students will see the actual diagnosis of the virtual child and the basis for that decision. In addition, corrective feedback would be available from the faculty for those students with low accuracy rates to allow them to discuss the process of making a diagnosis.

Conclusion

Our hope is to continue to expand this project as new technological resources become available. One option is to incorporate multimedia elements into the diagnostic process. For example, instead of providing the classroom observation results, the users could watch an animation or a video of a child in a classroom. This would allow the students to practice the skills necessary for conducting an observation that is a skill also taught in this course. Additional refinements could include the use of artificial intelligence to help guide the parent and teacher interviews.

Users could ask their own questions and the program would respond with appropriate answers. Clearly, the possibilities are endless as long as we remain open to learning and engaging new technologies as they emerge.

Another option is to adapt this teaching tool for use in other disciplines within psychology that practice diagnosis. For example, adult case studies could be used in the training of clinical psychology students. Similarly, it could be used with educational diagnosticians or social workers as a training tool. While the authors are unaware of the use of similar projects underway at other universities, this is clearly a portable technology. In summary, while this teaching tool in still a nascent technology, it shows considerable promise to fill an educational gap that exists in both School Psychology and other clinical fields.

References

Daly, E. J., Lentz, F. E., & Boyer, J. (1996). The instructional hierarchy: A conceptual model for understanding the effective components of reading interventions. *School Psychology Quarterly*, *11*, 369-386.

Daly, E. J., & Martens, B. K. (1994). A comparison of three interventions for increasing oral reading performance: Application of the instructional hierarchy. *Journal of Applied Behavior Analysis*, *27*, 459-469.

Fagan, T. K., & Wise, P. S. (2000). *School Psychology: Past, Present, and Future*. Bethesda, MD: National Association of School Psychologists.

Fischetti, B. A. (2000, September). Psychological evaluation time and components: Connecticut suburban practices. *NASP Communiqué*, *29*. Retrieved June 4, 2003, from http://www.nasponline.org/publications/cq291PsychEval.html.

Flanagan, D.P., & Ortiz, S.O. (2001). *Essentials of cross-battery assessment*. New York: John Wily & Sons, Inc.

Haring, N. G., Lovitt, T. C., Eaton, M. D., & Hansen, C. L. (1978). *The fourth R: Research in the classroom*. Columbus, OH: Merrill.

Issenberg, S. B., McGaghie, W. C., Hart, I. R., Mayer, J. W., Felner, J. M., Petrusa, E. R., et al. (1999). Simulation technology for health care professional skills training and assessment. *Journal of the American Medical Association*, *282*, 861-866.

Johnson, C. K. (1990). *Education and experience as determinants of diagnostic decision-making in school psychology trainees.* Unpublished dissertation, Texas Women's University, Denton, Texas.

Nathan, P. E. (1967). *Cues, decisions, and diagnosis: A system-analytic approach to the diagnosis of psychopathology*. New York: Academic Press.

Papa, F. J., Aldrich, D., & Schumaker, R. E. (1999). A new look at the diagnostic reasoning process. *Academic Medicine,* 74(10), S16-S18.

National Association of School Psychologists. (2002, Sept. 2002). *A career in school psychology*. Retrieved October 1, 2002, from http://www.nasponline.org/about_nasp/careerfaq.html

Thomas, M. D., O'Connor, F. W., Albert, M. L., Boutain, D., & Brandt, P. A. (2001). Case-based teaching and learning experience. *Issues in Mental Health Nursing, 22*, 517-531.

Tillman, T. C., Fiorello, C. A., & McGrath, M. C. (2001, Spring). The virtual student: Using technology to facilitate learning of the diagnostic process. *Trainer's Forum: Periodical of the Trainers of School Psychologists, 19*, 6-7.

Voytovich, A. E., Rippey, R. M., & Jue, D. (1986). Diagnostic reasoning in the more the problem patient. *Evaluation and the Health Professions, 9*(1), 9-102.

Training Techno-riffic Teachers

Jacqueline Hansen
Murray State University

Abstract

The author explores the technological revolution's impact on American pedagogical practices. Throughout history, instructional innovations have first met resistance and then acceptance by classroom teachers. The advent of instructional computer technology has added a challenging dimension. To facilitate the integration of computer technology into America's classrooms, the federal government passed the Telecommunications Reform Act in 1996. This act spawned the Technology Literacy Challenge, which provided equal access to technology, the Internet, and quality educational software for all American classrooms. This initiative also promoted training America's teachers to become technological proficient. Concurrently, multiple national agencies cooperatively created national technology teaching standards to promote teachers' technological proficiency. These standards have been adopted by the NCATE. The author details ways that she develops pre-service and in-service teachers' technological proficiencies by designing and implementing lessons incorporating the effective use of e-mail, *Blackboard, Kidspiration, Inspiration, Microsoft Word,* electronic portfolios and interactive satellite transmission.

"Our country was built on a simple value that we have an obligation to pass better lives and better opportunities on to the next generation. Education is the way we make that promise real. Today, at the dawn of a new century, in the middle of an information and communications revolution, education depends upon computers. If we make an opportunity to every student, a fact in the world of modems and megabytes, we can go a long way toward making the American Dream a reality for every student. Not virtual reality -- reality for every student."

President Clinton

Introduction

The launching of Sputnik in 1957 propelled our nation into a race-for-space that eventually led us into cyberspace. In response to the Russians' successful satellite send-off, the Federal Government passed the National Defense in Education Act in 1958, promoting math and science education in American schools. From 1959-1962, $70 million a year was provided to purchase technology hardware for educational use (Kauchak, Eggen, & Carter, 2002). Chalkboards gave way to whiteboards then SMART Boards. Black-and-white televisions, once used only for viewing special events such as the Apollo launches, began broadcasting a bevy of educational programs, ranging from Letter People to Channel One. Room-size computer systems designed to fly man to the moon became encapsulated into laptops. Traveling computer workstations multiplied into in-class computer labs.

How have teachers reacted to the integration of instructional technology into their classrooms? How can professors teach tomorrow's teachers the skills they will need to become technologically proficient professionals? This chapter explores the historic and modern impact of technological innovations on American classrooms and ways that professors can integrate computer technology into their teacher preparation programs.

In the purest sense, instructional technology involves any tools teachers use to assist in the transmission of information to their students, ranging from hi-liters to DVD players. Even with all of these available resources, many teachers have been reluctant to use instructional technology in their classrooms.

Technological innovations have been met with reluctance then acceptance throughout history. Teachers balked at chalk in the early 1800s. Professor Claude Crozet, who taught engineering at West Point in 1817, used a chalkboard to provide a visual aid while teaching descriptive geometry. "With the introduction of the chalkboard, teachers wondered, Could they learn to use this new chalk technology efficiently and how would this fresh innovation affect the curriculum?" (Sadker & Sadker, 2003, p. 520). Because some teachers did not want to use this "modern device," William Alcott and Josiah Bumstead wrote tutorials on how to use blackboards in the classroom. After initial reluctance, reformers eagerly embraced the use of chalkboards, so that by the early 1840s a chalkboard appeared in almost every northeastern American classroom. Now chalkboards (or modern whiteboards) are a traditional part of every classroom (Dockterman, 2002; Kauchak et al., 2002)!

In the 1860s many educators opposed the inclusion of photographs and illustrations in textbooks, stating that their usage would dilute the text's academic purpose. Eventually these feelings changed, however, so that by the time Thomas Edison touted the recently-invented motion picture as an educational tool in the 20th century, his technology was met with applause (Withrow, 1997; Dockterman, 2002).

It appears that, with training and experience, teachers' reluctance to embrace educational technology can be overcome. Does this hold true for modern-day teachers? Many of today's teachers have overcome their reluctance to use instructional technology such as whiteboards, illustrated texts, and videotapes. Now they face the new challenge of using computer technology effectively in their classrooms.

The total number of computers in K-12 American classrooms was estimated to reach 5.8 million during 1995, or one computer for every nine students. However, researchers have found that these computers are usually used for rudimentary drill and word-processing tasks. Fewer than 10 percent of teachers feel confident using presentation software, electronic networking, or problem-solving applications. Evidently the majority of teachers are not maximizing the potential power of instructional technology (The Office of Technology Assessment, 1995, as cited in Abdal-Haqq, 1995).

Even though computer technology is available in today's classrooms, not all practicing teachers are using it! Larry Cuban (2000) discovered that less than 10 percent of teachers used computers at least once a week, 20 to 30 percent used computers once a month, and over half of the teachers did not use computers in their classrooms at all! What causes teachers to be reluctant cyber-citizens? Do they fear that computer technology is an educational cyborg that will terminate traditional teaching techniques? Are they afraid to admit that their students might be more techno-savvy than they are? Herein lies the challenge. How can university professors turn computer curmudgeons into techno-riffic teachers?

The Four Pillars of Technology

In 1996, teachers' efforts to integrate computer technology were profoundly impacted by the Telecommunications Reform Act. This educational initiative provided a Technology Literacy Challenge Fund of $2 billion over a five-year period to ensure equal access to computer technology for all American students. Furthermore, the Telecommunications Reform Act also promoted distance learning, math and science instructional technology, and ongoing research into effective uses of educational technology (Kauchak et al., 2002; "Getting America," 1996).

The Technology Literacy Challenge focuses upon four pillars of technology. First, *all* students need to have access to computer technology. Parkay and Stanford (2001) stated "educational technology has a significant positive impact on student achievement" (p. 431). Because using technology increases student achievement, all children must have access to technology to reach their full potential and to develop technology skills necessary for the job market of the 21st century (Fielstein & Phelps, 2001; "Getting America," 1996; and Goldman, 1999; and Withrow, 1997).

Although American schools are designed to minimize socioeconomic and cultural differences by providing an equitable educational opportunity for all children, there has been a growing digital divide between cultures and socioeconomic groups in America. Fifty-four percent of white children have home computers, compared to African Americans (21%) and Hispanics (19%). Eighty percent of students living in households with incomes of $75,000 or more have home computers while only 16% of children living in low-socioeconomic conditions have access (Kauchak et al., 2002).

The second challenge of this reform effort was to provide Internet access for all schools. In 1995, President Clinton created the National Information Infrastructure to connect all schools, libraries, hospitals, and law enforcement agencies to the Internet. Between 1994 and 2000, the number of public schools in the United States with Internet access increased from 35 to 98 percent! The Federal Government made this dramatic increase possible by allocating more than $5.8 billion through Education rate (E-rate), a program providing discounts for telecommunications services and wiring to schools and libraries. Now almost all

American schools are online (Cattagni & Ferris, 2001; Parkay & Stanford, 2001; and Sadker & Sadker, 2003).

With computer hardware and Internet access in place, districts need to procure quality educational software. Although using computers for additional drill and practice of basic skills is important, teachers need to be mindful that computers also can be used to develop students' critical thinking and problem-solving skills (Kauchak, Eggen, & Carter, 2002; and Sadker & Sadker, 2003). The International Society for Technology in Education (ISTE) has promoted this concept by working with teachers throughout the country to identify quality educational software (ISTE, 2000).

Finally, teachers need to be trained to use modern technology in their classrooms. The nation's teaching force is rapidly changing. In 1991, about 20% of the 2.6 million public school teachers were 50 or older. Approximately 60% of first-time teachers in 1991-1992 were recent college graduates (Rand Organization, n.d.). With the multitude of retiring teachers, the increased number of students, and the demand for decreased class sizes, the federal government estimates that 2.2 million new teachers will be hired over the next decade (Milken Family Foundation, 2003). Today's teachers range from being technological pathfinders to slow adaptors, to technophobes, to digitally challenged (Sadker & Sadker, 2003). Only 20 percent of teachers express confidence in their ability to use instructional technology (National Center for Educational Statistics, 1999).

School districts invest money and time to train teachers to use instructional technology in their classrooms. But do they spend enough money? The National Education Association recommends that up to 40% of technology spending should be spent on training teachers, instead of on hardware and software as it is now (Sadker & Sadker, 2003).

Do school districts invest enough time to train their teachers? Time is teachers' nemesis. Lack of time interferes with even the most enthusiastic educators' efforts to employ novel strategies in their classrooms. To implement newly learned instructional technology, teachers need more time to polish their proficiencies, collaborate with colleagues, and plan authentic applications that actively involve students. Achieving technological proficiency may take up to five years of training and coaching (Metcalf & Jolly, 2002; Sadker & Sadker, 2003; Rand Organization, n.d.).

Integrating technology throughout all of America's schools requires training of preservice and practicing teachers (Rand Organization, n.d.). Teachers need to begin to learn about the effective integration of instructional technology before they enter the classroom. During their first year, many beginning teachers focus on classroom management, curriculum, and their new role as teachers. They consider computer usage an "extra" in addition to the core curriculum (Strudler, N.B., McKinney, M.O., & Jones, W. P., 1999). Beginning teachers who do feel prepared to use instructional technology attribute their competence to required preservice technology courses, the degree to which instructional technology was used in methods courses, and the use of computers during the student teaching experience (Handler, 1993, as cited in Strudler, et al., 1999).

What technology proficiencies do teachers need to demonstrate? The ISTE, in conjunction with the U. S. Department of Education, NASA, the Milken Exchange on Education Technology, and Apple Computer, Incorporated, developed national education technology standards for students, administrators, and teachers. These teaching standards, adopted by the National Council for the Accreditation of Teacher Education, mandate formal technology training for teachers (ISTE, 2000; Kauchak, et al., 2002; Benson, 2000). The standards state that preservice and practicing teachers need to demonstrate an understanding of technology operations and concepts, plan and design effective learning environments incorporating technology, and apply technology to maximize and assess student learning (ISTE, 2000; Dockterman, 2002). Now, 42 states have adopted technology standards and require formal teacher training in technology (Kauchak et al., 2002). Kentucky is one of these states.

To address this perceived need in teacher education programs, Kentucky educators and legislators designed a series of New and Experienced Teacher Standards that detail what teachers should know and be able to implement in their classrooms. New Teacher Standard IX and Experienced Teacher Standard X directly address the use of instructional technology in the classroom. Both standards include identical verbiage: "The teacher uses technology to support instruction; access and manipulate data; enhance professional growth and productivity; communicate and collaborate with colleagues, parents, and the community; and conduct research." (Kentucky Education Professional Standards Board, p. 4).

Preservice and practicing teachers are evaluated according to their ability to (Kentucky Education Professional Standards Board, 1993, p. 4):

1. operate a multimedia computer and peripherals to install and use a variety of software.
2. use terminology related to computers and technology appropriately in written and verbal communication.
3. demonstrate knowledge of the use of technology in business, industry, and society.
4. demonstrate basic knowledge of computer/peripheral parts and attend to simple connections and installations.
5. create multimedia presentations using scanners, digital cameras, and video cameras.
6. use the computer to do word processing, create databases and spreadsheets, access electronic mail and the Internet, make presentations, and use other emerging technologies to enhance professional productivity and support instruction.
7. use computers and other technologies such as interactive instruction, audio/video conferencing, and other distance learning applications to enhance professional productivity and support instruction.
8. request and use appropriate assistive and adaptive devices for students with special needs.
9. design lessons that use technology to address diverse student needs and learning styles.
10. practice equitable and legal use of computers and technology in professional activities.
11. facilitate the lifelong learning of self and others through the use of technology.
12. explore, use, and evaluate technology resources: software, applications and other documentation.
13. apply research-based instructional practices that use computers and other technology.
14. use computers and other technology for individual, small group, and large group learning activities.
15. use technology to support multiple assessments of student learning.
16. instruct and supervise students in the ethical and legal use of technology.

To actualize this vision of effective implementation of instructional technology, Kentucky legislators created the Kentucky Academy of

Technology Education (KATE) in 1996. The purpose of this academy is to design professional development opportunities to improve the use of instructional technology. Although the KATE offices and personnel are located at Murray State University (MSU), their efforts outreach to all educational institutions and classrooms throughout the state. KATE experts work closely with MSU's College of Education faculty. KATE's mission is to encourage faculty and pre-service teachers to learn to integrate technology throughout the educational process. Because of Kentucky educators' and legislators' focus upon instructional technology, Kentucky students now rank among the top five states in the use of technology. (B. Nix, personal communication, August 29, 2002; KATE, 2002).

For the past several years, KATE has provided extensive training opportunities for the faculty and student teachers of the College of Education (COE) at Murray State University. Moreover, KATE experts worked closely with COE administrators in the design of the new education facility, Alexander Hall. Every classroom is equipped with two desktop computers, a SMART Board projection system, TV, overhead projector, and VCR. All faculty members have desktops in their offices and laptop computers to connect to SMART Boards (www.smarttech.com) for classroom presentations. Alexander Hall houses two computer classroom labs and a large student computer lab, with over 150 computers available for faculty and student use! Professional development opportunities are ongoing. Moreover, KATE has designated and trained a team of e-mentors, faculty "computer coaches" who support and scaffold the technological efforts of their colleagues. For their visionary leadership, the College of Education and KATE recently received the 2003 Distinguished Achievement Award from the International Society for Technology in Education ("MSU Claims," 2003).

Training Tomorrow's Teachers

KATE e-mentors, or faculty "computer coaches," work continuously to promote the effective integration of instructional technology into the College of Education's course offerings. E-mentors are also expected to model best instructional technology practices in their classrooms. As a member of the College of Education faculty, I have gained the training and experience to become a KATE e-mentor who embraces the use of instructional technology in the delivery of my courses. How do I use instructional technology to train pre-service and in-service teachers to become technologically proficient in their classrooms?

As I learn new technology, I teach my students what I know, and they teach me! First, I establish an atmosphere of classroom community where people feel comfortable to admit their ignorance and their expertise so that we can help one another learn. I model lifelong learning by trying new technology in my classroom and actively seeking new ideas from my students. I also maintain a consistently enthusiastic and positive attitude about the use of instructional technology, even when technical difficulties arise!

Next, I determine where my students are on their technological journeys. A 2003 study by the Milken Family Foundation found that over 70% of teacher preparation programs include a required technology course. As part of the MSU Core Studies, all students are required to take CSC 199 *Introduction to Information Technology.* In this course they learn how to operate Power Point, data bases, spreadsheets, and word processors. I extend and enrich these experiences by integrating technology throughout my course delivery and assignments. As part of the NCATE accreditation continuous assessment process, the COE works with KATE to document our students' technological journeys. The KATE survey instrument (http://www.hprtec.org/profiler) is administered at key checkpoints at the beginning, middle, and end of preservice teachers' academic careers. Results from this longitudinal study are used to refine the teacher preparation program and to identify areas of professional development for faculty and students. Technological targets have been established for each of the core education courses to ensure that all students have the opportunity to develop their proficiencies.

I integrate technology throughout the design and delivery of diverse undergraduate and graduate courses. Students are first introduced to the importance of using instructional technology in *Issues and Practices of American Education*, an introductory level course designed for students who are interested in pursuing careers in education or becoming more knowledgeable about the American educational system. Students learn how to use technology in the design and delivery of instruction and the evaluation of students' learning in multiple undergraduate and graduate courses, including: *Strategies of Teaching*, *Teaching Early Elementary*

Language Arts, *Teaching Early Elementary Mathematics*, *Evaluation and Measurement in Education*, and *Readings and Research in Children's Literature*. Student teachers also integrate technology throughout their 14-week student teaching experiences.

Blackboard

Experts from the MSU Center for Teaching, Learning and Technology (CTLT) trained me to use Blackboard (www.blackboard.com) for all of my courses. This outstanding computer program enables students to retrieve course materials and correspond via the Internet. When students first log in, they encounter an announcement page where I post "Mama Hansen messages" and the weekly calendar. Students click on a series of hyperlinks to access the course syllabus, calendar, assignments, faculty information, and related websites. Under "student tools," students update their e-mail addresses, see their grades, create home pages, and organize their lives with task-lists, journals, and customized calendars. The Blackboard communications component allows us to e-mail individuals, groups, or the entire class. Students may submit assignments with attached files or by using the digital drop box. Students in the *Readings and Research in Children's Literature* course use the discussion board feature to conduct threaded discussions about selected topics and to share annotated bibliographies of outstanding children's books.

The Blackboard control panel allows me to customize the program for each course. I work with CTLT and the registrar to create or replicate Blackboard sites for each new semester. I post information through folders and hypertext links. Students can view their ongoing grades in an Excel-style program. All course materials can be archived for future use.

To ensure consistent, quality delivery of our core curriculum, I give other course instructors reader-only access to my Blackboard site. I also create a cohort group of instructors so that they can use this site as a forum to share course comments and concerns.

Because accessing Blackboard is the only way my students can get their assignments, they rapidly become proficient users. As with any program, there are some occasional difficulties, however. At times students are unable to access Blackboard because the server is overloaded or malfunctioning. Our university and college are taking measures to expand servers and remediate that problem. Files stored in the digital dropbox need to be saved, opened, then resaved with comments and sent to students. To expedite the grading process, I ask students to send their assignments via e-mail instead.

E-Mail

E-mail, or electronic mail, is an increasingly popular way to correspond (Kauchak et al., 2002). In all of my classes, students use e-mail to correspond with me, and their cooperative teammates and classmates. Those who are unable to attend class are expected to e-mail or voice-mail me *before* class begins. Students use e-mail to correspond with me and to submit late assignments.

I used to require students to e-mail all their assignments. I was able to intersperse commentary into their efforts by responding in all caps of a different color or by saving their files and inserting pop-up commentary balloons using Microsoft Word. I soon discovered that I was spending most evenings in front of a computer monitor, however, which negatively impacted my quality family time. Therefore, students now give me paper copies that I can check while sitting on the couch, communing with my family.

I also incorporate e-mail into class projects. One of the major goals of *Issues and Practices of American Education* is to acquaint students with the personal qualities, knowledge, and skills necessary to become an effective educator. Students conduct e-mail interviews with master teachers or administrators from across the country. They send me carbon copies of the e-mails and responses. Students then write modern educational leader reports to share with their classmates. This is an excellent way to bring a multitude of master educators into our Kentucky classroom!

During the *Teaching Elementary Mathematics* and *Teaching Early Elementary Language Arts* courses, students teach multiple lessons in regional elementary classrooms. To ensure an effective educational experience, they correspond with cooperating teachers via e-mail.

I also introduce students to multiple means of using technology in their future classrooms. An increasingly popular program is Epals, a site that allows students from over 100 countries and 13,600 classrooms to become electronic pen-pals. Participants report that using Epals improves students' written communication skills, builds shy learners' confidence, and improves all students' computer skills (Kauchak et al., 2002).

I personalize this concept by having my *Teaching Early Elementary Language Arts* students correspond with local key-pals about stories they have written, books they have read, or other topics of interest. Each semester, university and elementary students become e-mail partners. This activity helps university students learn more about the developmental stages and interests of elementary students. Elementary students are given an authentic audience for their writing efforts. This year I plan to extend this program by having the key-pals meet for a social occasion during the semester.

I correspond at least once a week with student teachers via e-mail to discuss their trials and tribulations and to arrange appointments for multiple formal observations. Student teachers also e-mail me a summarization of weekly activities, professional growth plans, and professional portfolio reflections.

Using e-mail has become an expedient way to correspond with colleagues and students. There are occasional difficulties, however. Because I use the Blackboard site to e-mail individuals or groups of students, they need current e-mail addresses. These addresses also need to be professionally appropriate. One student was unable to get a job working with children after school because his e-mail address was crude and school officials were unsure if that e-mail address was indicative of his moral character. Some students do not use spell-check or proper language mechanics when composing e-mail messages. That is why I have built in professional accountability by having students send me carbon copies of their correspondence to supervising teachers and key-pals. Sometimes the COE or MSU servers are down or students' inboxes are full, which limits access. I also need to be wary of viruses and attached files created by a non-compatible system (I use Microsoft Word which turns Word Perfect documents into "alien talk"). Also, because students oftentimes study throughout the night, they might e-mail questions to me at 3:00 AM and then become concerned because they have not received a response by 8:00 AM the next morning! Finally, sometimes e-mail simply disappears into cyberspace, so I encourage students to keep a copy of important papers on their hard drives just in case.

Distance Education

I especially depend upon e-mail to correspond with and receive assignments from my long-distance graduate students. Electronic mail and satellite technology has profoundly impacted educational delivery by expanding the virtual campus to include students who, because of geographic distances, might not be able to receive degrees in the traditional classroom setting.

I first became familiar with the power of e-mail and interactive satellite instruction while taking long-distance doctoral courses through the University of Nebraska-Lincoln. Distance education is "the 'e-quivalent' of correspondence courses" (Sadker & Sadker, 2003, p. 530). Qualified faculty members deliver quality courses to remote locations throughout a broad geographic area. This form of course delivery allows nontraditional and working students an opportunity to complete advanced degrees. In distance education courses, instructors conduct the class on-campus to a traditional group of students. Class activities are simultaneously transmitted to long-distance sites via two-way television/satellite equipment (Fielstein & Phelps, 2001; Kauchak et al., 2002; and Sadker & Sadker, 2003). MSU Instructors who teach distance education courses are trained use the Elmo and transmission equipment. Service technicians are also on-hand to assist in course transmission.

Readings and Research in Children's Literature is a distance education course for graduate students. Because this course is offered during the first summer session, sometimes teachers are still teaching elementary classes. To accommodate students' schedule, the first few class sessions are offered totally online via Blackboard and e-mail. This creates unique challenges. How do I contact all students taking the course to notify them of the class-meeting schedule? I work with the registrar and site coordinators to identify and contact students enrolled for the course. How can I create the all-important sense of classroom community at the outset of the course? Students introduce themselves using threaded discussions on Blackboard. In the future, I plan to have them also create mini-home pages with digital pictures to share on Blackboard.

Once we meet "live" via interactive satellite, we spend the first class session becoming better acquainted. Students work closely with their site technicians to learn how to operate the equipment for class communication and their group presentations. It took me awhile to become acquainted to this new teaching medium. I am a hands-on, walk-around teacher. Transmitting long distance requires me to be rather stationary. Because I believe in actively involving students in authentic activities, I do not want to be a satellite-transmitted "talking head." Therefore, I intersperse mini-lectures with independent and cooperative activities (just as I do when teaching traditional courses). I set up the transmission so that students and I can see other sites on a rotating basis. I create cooperative teams at each site. Sites hosting one or two students are teamed with students at other sites for cooperative activities. These teams keep their microphones turned on so they can communicate. Once I give students their team assignments, I turn off my local microphone and set the cameras to view their activities. People who need assistance notify me by writing a "contact us" memo on their Elmo. I post a "two-minute warning" sign on the Elmo to cue students that we will be reconvening. Students have responded very favorably to this active-learning approach.

Technology has improved since I took my first long distance course, so it has become rare for interrupted transmissions of the satellite signal. When this does happen, I fax lecture notes and group assignments to the site that is unable to receive the transmission. We also communicate by telephone.

Internet

In this information age, the Internet/Worldwide Web has become an international repository of resources. Accessing the web enhances students' and teachers' efforts to conduct research and develop quality lessons (Kauchak et al., 2002; Lan, 1999). Students can conduct research by accessing ERIC, Kentucky Virtual Library, and all MSU library materials online. They can view the COE website to see advising forms, schedules, course catalogs, faculty information, and teacher education materials (http://www.murraystate.edu/coe). The KATE website offers multiple resource materials including hotlinks to educational websites, tutorials for various educational computer programs, and student teacher information (http://coekate.murraystate.edu/kate/).

Each semester I give mini-lessons introducing relevant Internet sites and addressing issues on plagiarism and judicious use of quality websites. Each week, students in the *Issues and Practices of American Education* course access the web to find additional information and/or news items related to that week's topic. They share these "tech connections" with classmates the date the topic is presented in class.

Research on the Internet is a common academic activity for students (Parkay & Stanford, 2001). All of my students have a required research component in their coursework. For teachers to continue to grow personally and professionally and to truly meet the needs of their students, they need to be aware of current educational trends and historic educational practices. Students conduct cooperative inquiry projects on approved topics related to the course objectives. They conduct research using traditional reference materials as well as the Internet. Each student documents his/her research efforts by submitting summaries, note cards, or highlighted copies of the materials. Students also submit annotated bibliographies of resource materials and a typed list of key points. Then, instead of creating a research paper, student teams create a formal presentation for their classmates. They use a publishing program to create handouts or brochures of key information, make a poster, and create a multimedia PowerPoint presentation. This approach minimizes the possibility of plagiarism. In traditional research papers, students can copy-and-paste information or buy entire papers on the Internet. When students summarize and document research efforts to share with their classmates they replicate the process that practicing teachers use when sharing information with colleagues at their school-sites. Furthermore, through this approach, students have yet another opportunity to speak publicly and to work in teams.

The Internet is an amazing resource for lesson-planning ideas. According to the U.S. Department of Education's Teaching, Learning, and Computer survey, up to 68 percent of teachers use the Internet as a resource when planning their lessons (Parkay & Stanford, 2001). Students and I both use the Internet to enrich and extend required curriculum materials. Students taking my methods courses are required to find and share (via e-mail) the annotated descriptions of ten exemplary educational websites. I also provide outstanding lesson plan links on Blackboard.

I encourage all of my students to browse the net for lesson ideas. Students are welcome to use these ideas as long as they modify the lessons to the point where they "make them their own." Furthermore, they must give credit to the original source and author. When developing or implementing instructional units, or a series of interrelated lessons, students are required to include at least one technology activity. Students in my methods classes often use the teaching lab or portable laptop labs to teach lessons using specialized software or Internet sites. Student teachers must use technology to design and/or implement each of the lessons that I formally observe. All lessons are written using the lesson plan format required by the state of Kentucky. Lesson plan templates are available on my Blackboard site and through hotlinks to the Kentucky Department of Education. Students are also welcome to use Lesson Plan Creator to write their lessons.

Presentations

Using technology enables students to produce quality, professional products. All student assignments are typed using word processing programs. Researchers state that students enjoy composing on computers because of ease of editing text, getting information on paper, combining ideas, checking spelling/grammar, and creating legible products (Kauchak et al., 2002).

Students also use instructional technology for their formal presentations. I train students taking the *Strategies of Teaching* and *Teaching Early Elementary Mathematics/Language Arts* courses to use Kidspiration, Inspiration, and PowerPoint. Kidspiration and Inspiration (www.inspiration.com) are computer graphing programs that allow students to create concept webs with interrelated text. Kidspiration, Inspiration, and Power Point provide a visual means of communicating complex information and the interrelatedness of ideas to students.

First, I model best practices by using this presentation software in my lesson delivery. Then, I reserve a faculty teaching lab and systematically train students to use these programs for instructional purposes. We discuss reasons for using presentation software before I formally introduce the program features. Then cooperative teams create authentic products demonstrating their new knowledge.

Students demonstrate their individual competencies by integrating the programs into their lessons and research presentations. In *Strategies of Teaching,* students create webs depicting all of the things teachers must consider when planning to teach. They revisit these webs at the end of the semester and reflect upon what they have learned about the teaching process. Some *Strategies of Teaching* students also use Kidspiration/Inspiration to record student responses during brainstorming sessions. Students in all of my methods courses use Kidspiration/Inspiration to graphically represent their instructional unit designs. In *Teaching Early Elementary Language Arts*, students use Kidspiration or Microsoft drawing tools to create detailed designs of language-rich classrooms.

Because technology involves much more than computers, I also train students to use the whiteboard, overhead projector, and video camcorders in their instructional efforts. In *Strategies of Teaching*, students' microteaching efforts are videotaped so that they can view and adjust their nonverbal behaviors and instructional delivery.

Although training students to use technology takes cherished instructional time, the results are worth the effort! To accommodate the varying needs of students, people who are already proficient become peer-mentors. Struggling learners access tutorials and receive additional one-on-one instruction from me. A word of caution—be aware of which version of the program students are using outside the classroom and find out if it is compatible with the version of the software package installed in the presentation computers used in the classroom. Modernized versions sometimes give the option of saving the new graphics in an older format compatible with the software on the presentation computer. The best way to avoid translation difficulties is to update the presentation computer software and/or to encourage students to use the same version when creating their graphic designs. Require students to test well in advance their version on the presentation computer to make sure it works.

Student Portfolios

Students also develop their technological proficiency by constructing electronic portfolios. To become certified in the Commonwealth of Kentucky, beginning teachers must demonstrate their progress towards mastering each of these Kentucky New Teacher Standards (Kentucky Education Professional Standards Board, 1993, pp. 1-4):

I. Designs/Plans Instruction
II. Creates/Maintains Learning Climates
III. Implements/Manages Instruction
IV. Assesses and Communicates Learning Results
V. Reflects/Evaluates Teaching/Learning
VI. Collaborates with Colleagues/Parents/Others
VII. Engages in Professional Development

VIII. Knowledge of Content

IX. Implementation of Technology

First-year teachers demonstrate their teaching abilities through professional portfolios. They include an artifact and statement reflecting upon their proficiency for each of the nine standards. All COE courses are designed to help students develop proficiencies in each of these standards. Recommended artifacts and reflection papers are required for each course. Furthermore, to assist students' efforts, they are required to develop formative and student teaching portfolios with artifacts and reflective statements demonstrating their progress towards meeting the standards.

The portfolio process begins during their initial education course, *Issues and Practices of American Education.* During this course, the College of Education NCATE Coordinator makes a special presentation about the need and benefits of developing professional portfolios. Then instructors work closely with the KATE experts to help students create electronic formative portfolios stored on the COE server. Students are encouraged to save electronic copies of artifacts and reflective statements in these folders as they progress throughout their academic coursework at MSU. They are also trained to create Web pages containing the same information. Students' progress on portfolios is checked at regular intervals by advisors and instructors throughout the students' coursework. Then, when students reach their final semester, they select one artifact and reflection that best demonstrates their professional progress in each of the nine New Teacher Standards. Two faculty members formally evaluate student teaching portfolios.

The formative and e-portfolio process has become so successful that it is now a required component of *Issues and Practices of American Education.* All instructors of this course have been formally trained by the KATE experts. These experts have facilitated teachers' and students' efforts by providing detailed instructions located on the COE homepage at http://coekate.murraystate.edu/portfolio.

Because formative and e-portfolios were piloted in selected classrooms, not all students who took *Issues and Practices of American Education* in the past have created formative portfolios. Therefore, faculty members are working to identify other core courses, such as *Strategies of Teaching,* where students can begin the formative portfolio process. All formative portfolios must be underway before students can be admitted to the Teacher Education Program.

Your Technology Literacy Challenge

Even though I grew up during the Sputnik era, I admit that I am a recovering technophobe. We recently upgraded our home stereo system and I still do not have the "remotest" idea what some of the buttons are on the remote control! How have I become a technologically proficient professor? By modeling the importance of lifelong learning and believing in my ability to equip my students with the skills they will need to become techno-riffic teachers.

Is using instructional technology expensive? Yes. Does it require a great deal of time and effort? Yes. But helping our students become global citizens who are able to impact the lives of children for generations to come makes this journey worthwhile. And so I challenge you to launch into a lifetime of learning how to enthusiastically and actively integrate instructional technology into your college classroom. Begin your own personal Tech Trek. Dare to join the cyberspace race and go where no one has ever gone before. The world is literally only a keystroke away. You CAN do it.

References

Abdal-Haqq, I. (1995, Nov.). *Infusing technology into preservice teacher education.* Retrieved July 10, 2003, from http://ericae.net/ericdb/ED389699.htm.

Benson, S.J. (2000). *Preparing prospective teachers for technology integration.* Retrieved July 11, 2003, from http://pt3.nmsu.edu/educ621/sharon1.html.

Cattagni, A., & Ferris. E. (2001, May). *Internet access in U.S. public schools and classrooms, 1994-2000.* Washington, D.C.: National Center for Education Statistics. Retrieved July 11, 2003, from http://nces.ed.gov/pubs2001/InternetAccess/2.asp.

Cuban, L. (2001). *Oversold and underused: Computers in the classrooms, 1980-2000.* Cambridge, MA: Harvard University Press.

Dockterman, D. A. (2002). *Weaving technology into your teaching.* USA: Tom Snyder Productions, Inc.

Fielstein, L., & Phelps, P. (2001). Integrating computer technology. In D. Alpert (Ed.), *Introduction to teaching: Rewards and realities*, (pp. 287-308). Belmont, CA: Wadsworth/Thomson Learning.

Getting America's Students Ready for the 21st Century: (June, 1996). *Meeting the technology literacy challenge*. Retrieved July 11, 2003, from http://www.ed.gov/Technology/Plan/NatTechPlan/priority.html

Goldman, S.R., Williams, S. M., Sherwood, R. D., & Hasselbring, T.S. (1999). Technology for teaching and learning and understanding. In J.M. Cooper, (Ed.), *Classroom teaching skills* (6th ed.). (pp. 181-219). USA: Houghton Mifflin Company.

International Society for Technology in Education. (2000). *National educational technology standards for students connecting curriculum and technology*. USA: ISTE and US Department of Education.

Kauchak, D., Eggen, P., & Carter, C. (2002). Technology in American schools. In J. W. Johnston (Ed.). *Introduction to teaching: Becoming a professional* (pp. 374-406). Upper Saddle River, NJ: Pearson Education Inc.

Kentucky Education Professional Standards Board (1993). *New teacher standards for preparation and certification*. Frankfort, KY: Education Professional Standards Board.

Lan, J. (1999). *The impact of internet-based instruction on teacher education: The paradigm shift.* Retrieved July 17,2003, from http://www.ed.uab.edu/jlan/aacte99/aacte99.htm.

Metcalf, T., & Jolly, D. (2002, January). *Intel teach to the future: A partnership for technology staff development.* Paper presented at the 9th Annual Distance Education Conference. Retrieved July 14, 2003, from http://www.cdlr.tamu.edu/dec_2002/Proceedings/Metcalf.pdf.

Milken Family Foundation. (2003). *Information technology underused in teacher education.* Retrieved July 11, 2003, from http://www.mff.org/edtech/article.taf?_function=detail&Content_uid1=131.

MSU claims award from ISTE. (July 16, 2003). *Murray Ledger & Times*, p. 12A.

National Center for Educational Statistics (1999). *Teacher quality: A report on the preparation and qualifications of public school teachers*. Washington, DC: U.S. Department of Education. Retrieved July 11, 2003, from http://nces.ed.gov/surveys/frss/publications/1999080/

National Center for Education Statistics. (2002). *A report on qualifications and preparation of public school teachers*. Retrieved July 11, 2003, from http://nces.ed.gov/surveys/frss/publications/1999080/

Parkay, F. W. & Stanford, B. H. (2001). Teaching with technology. In P.A. Smith (Ed.) *Becoming a teacher* (5th ed.), (pp. 412-453). Needham Heights, MA: Allyn & Bacon

Rand Organization. (n.d.). Challenges of creating a nation of technology-enabled schools. Retrieved July 9, 2003, from http://www.rand.org/publications/MR/MR682/ed_ch4.html#RTFToC7.

Sadker, M.P., & Sadker, D. M. (2003). Technology in education. In Cara Harvey (Ed.), *Teachers, schools, and society* (6th ed.), (pp. 516-549). New York: McGraw Hill.

Strudler, N.B., McKinney, M.O., & Jones, W. P. (1999). First-year teachers' use of technology: Preparation, expectations and realities. *Journal of Technology and Teacher Education*, 7(2), 115-129. Retrieved July 11, 2003, from http://www.aace.org/dl/files/JTATE/JTA.

Withrow, F. (June 1997). Technology in education and the next twenty-five years. *Technological Horizons in Education Journal Online*. Retrieved July 11, 2003, from http://www.thejournal.com/magazine/vault/articleprintvesion.cfm?aid=1680.

Utilizing WebQuests in a Preservice Teacher Education College Course

Bob Perkins
College of Charleston

Brenda Roughton
James Island Elementary

Abstract

The investigators of this project include a college professor within the school of education and a third grade teacher. The candidates are in a preservice teacher education program at the College of Charleston. As a class assignment for Technology in Education course, each candidate prepared a WebQuest using topics suggested by the third grade teacher. The authors selected one project based on quality, the degree it addressed third grade state standards, and trueness to the WebQuest concept. The third grade students were transported to the College to participate in the selected WebQuest. Each candidate facilitated a group of three to four third grade students. As a follow up to this assignment, both the candidates and elementary students were surveyed for the positive and negative aspects of this project.

Introduction

WebQuests involve having students use the web to complete a task or solve a problem. The teacher creates the WebQuest by defining the problem to be solved and the steps involved, and providing resources for the students. In this project, the teachers are teacher candidates (from here on called candidates) in a preservice teacher education program. Each of the candidates prepared a WebQuest as a classroom project for their Technology in Education class. Out of those projects, one was selected based on a set of criteria: holding to the requirements of a WebQuest, interest to third graders, and readability. An elementary teacher brought her third grade students to the College to use the WebQuest that was selected. We surveyed both the candidates and elementary students to see what they felt were the positive and negative aspects of this project.

Technology and Constructivism

An important concept in instructional technology is constructivism. In constructivist theory, students "construct" their own understanding of the world (Tiene & Ingram, 2001). While other theories stress the importance of the teacher, constructivism stresses the learner (Tiene & Ingram). Knowledge is created, by relating what is encountered to what is already known (Jonassen, Peck & Wilson, 1999). Jonassen, Peck & Wilson (p. 3-5) make the following observations regarding knowledge:

- Constructivists believe that knowledge is constructed, not transmitted.
- Knowledge construction results from activity, so knowledge is embedded in activity.
- Knowledge is anchored in and indexed by the context in which the learning activity occurs.
- Meaning is in the mind of the knower. Therefore, there are multiple perspectives on the world.
- Meaning making is prompted by a problem, question, confusion, disagreement, or dissonance (a need or desire to know) and so involves personal ownership of that problem.
- Knowledge-building requires articulation, expression, or representation of what is learned (meaning that is constructed).

According to Tiene and Ingram (2001), certain principles are part of constructivist learning. First, learning must take place in problem-solving situations. Second, scaffolding, or someone helping the learner in the development of new learning, is supported through the learning experience. Eventually, these supports are removed so that the learner is able to connect previous experience with the new context on their own. Problem-based learning is an example of constructivism in use (Tiene & Ingram, 2001). "Engaging students in solving real, open-ended problems is a very powerful strategy for learning" (Brooks, Nolan & Gallagher, 2001, p. 71). Being able to use the Internet to connect classrooms to information is an example of technology's relationship to constructivist theory (Tiene & Ingram, 2001).

What Are WebQuests?

WebQuest, a term coined by Dodge and March in 1995, is a technology tool for educational use of the Internet and World Wide Web. WebQuests incorporate the constructivist model of teaching into using the World Wide Web. "A WebQuest is an inquiry-oriented activity in which some or all of the information that learners interact with comes from resources on the Internet, optionally supplemented with videoconferencing" (Dodge, 1997, para. 2). WebQuests are becoming quite popular in K-12 education but little research has been published in educational journals concerning their effectiveness. WebQuests are currently being used by teachers to facilitate web searches and for project-based learning (DeVincentis & King, 2000).

According to Dodge and March's (1997) web site at San Diego State University, the following are critical attributes of a WebQuest:

> WebQuests of either short or long duration are deliberately designed to make the best use of a learner's time. There are questionable educational benefits in having learners surfing the net without a clear task in mind, and most schools must ration student connect time severely. To achieve that efficiency and clarity of purpose, WebQuests should contain at least the following parts:
>
> 1. An **introduction** that sets the stage and provides some background information.
> 2. A **task** that is doable and interesting.
> 3. A set of **information sources** needed to complete the task. Many (though not necessarily all) of the resources are embedded in the WebQuest document itself as anchors pointing to information on the World Wide Web. Information sources might include web documents, experts available via e-mail or real-time conferencing, searchable databases on the net, and books and other documents physically available in the learner's setting. Because pointers to resources are included, the learner is not left to wander through webspace completely adrift.
> 4. A description of the **process** the learners should go through in accomplishing the task. The process should be broken out into clearly described steps.
> 5. Some **guidance** on how to organize the information acquired. This can take the form of guiding questions, or directions to complete organizational frameworks such as timelines, concept maps, or cause-and-effect diagrams as described by Marzano (1988, 1992) and Clarke (1990).

6. A **conclusion** that brings closure to the quest, reminds the learners about what they've learned, and perhaps encourages them to extend the experience into other domains.

From a theoretical perspective, WebQuests are recognized as incorporating a constructivist approach to educating students (Dutt-Doner, Wilmer, Stevens & Hartmann, 2000).

Issues Related to the Use of WebQuests

There are many good reasons for using WebQuests when students are accessing the web. There is concern with students having free access to the Internet. WebQuests address many of those concerns (Summerville, 2000; and Yoder, 1999). There are many controversial sites on the Internet, but because web sites used in a WebQuest have already been researched, it is not necessary for students to use search engines that may lead to inappropriate sites. Another problem with using search engines is the large amount of Internet information that could be potentially found, but again, the web sites used as resources in the WebQuests are selected by the teacher and provided as links in the Resource page of the WebQuest.

Creators of WebQuests must take into consideration the many facets of the educational uses of the web. Compiling a list of links is not enough to create a WebQuest. Dodge and March (1997) have defined what constitutes a WebQuest, but there are other considerations such as proper web page design. Instructional strategies to prevent learners from losing their places in the WebQuest need to be built in (Lin & Hsieh, 2001), and providing learner control of lessons must be approached cautiously (Lin & Hsieh, 2001). It is critical to provide a way for students to find their way back to a home page. Icons should be used to assist early learners in locating the home page.

Summerville (2000) stated that teachers must also be aware of other problems that exist within WebQuests. Students may still get to inappropriate sites, accidentally or on purpose. Teachers may want to have a plan to prevent these accidents such as using filtering technology. Developing WebQuests also takes time, and the amount depends on the teacher's web page development skills and the length of time students are expected to spend conducting the WebQuest. The last problem cited is "dead links," which are links that may have worked, but do not work at the time the WebQuest is used by students. Since the web is constantly changing, teachers will need to check their WebQuest links before students attempt to use them.

The use of WebQuests is starting to creep into research literature. Dutt-Doner et. al (2000) used a WebQuest to incorporate the state of Maine Learning Results and technology standards into their middle school curriculum. Their reasons for using a WebQuest were to integrate technology into their curriculum, use the learners' time to focus on using information rather than finding it, and taking students to more in depth levels of analysis, synthesis and evaluation. Problems encountered in their project included: students obtaining inaccurate information, teachers trying to manage too many students on the computers at one time, and logistical problems for first time "WebQuesters."

A WebQuest in a Preservice Teacher Education Program

The investigators of this project designed a WebQuest learning experience to bring together two unique groups of students. Teacher candidates, enrolled in an undergraduate teacher education program, were brought together with third grade students at a local elementary school. The goals of this project were to explore: (1) whether third grade students learn from WebQuests, (2) whether teacher candidates can create effective WebQuests. The anticipated results were that the candidates would develop effective WebQuests that meet state curriculum standards to produce student learning and that the third graders would learn the objectives related to a thematic unit that they were involved within their classroom.

Teacher candidates are typically in the first semester of their junior year and the first year of their education courses. At this level, they have had little experience working directly with children. Instructing candidates in using instructional technology without children provides a sterile learning experience. In the course, Technology for Teachers, candidates must complete two assignments that involve creating computerized lessons. One assignment uses PowerPoint and the other assignment is creating a WebQuest. In order to provide an optimal learning experience, teacher candidates need to have hands-on experience working with children. Candidates need to realize that lessons that look good on paper do not always work in real life. Having the candidates in this technology course create a WebQuest for a local third grade class, and bringing the students to the college was an excellent way of providing teacher candidates an opportunity to work with children.

The candidates were the developers and facilitators of the WebQuests. They were given an assignment to create a WebQuest based on three units that the third grade teacher was teaching: rocks, Christopher Columbus, and a language arts unit based on The Wreck of the Zephyr. Secondary and special education candidates in the class were given the option of choosing topics more relevant to their anticipated teaching level. All candidates had to incorporate the state curriculum standards into their WebQuest.

Instruction for the candidates started with an example WebQuest created by the instructor. Candidates completed the WebQuest to experience how students would interact with a WebQuest. The instructor wanted the candidates to experience the critical parts of a WebQuest and to understand that WebQuests are based on solving a real problem that does not already have a solution. Candidates were then introduced to sample WebQuests and instructed to consult Dodge's website for rules governing the quality of WebQuests. Students were taught how to create web pages using Netscape Composer. This program was chosen because it is freely available to all students. The last part of their instruction was how to transfer their web pages onto the college server. The students had access to the classroom computer lab as well as two student labs open for extended hours.

After the WebQuests were submitted, the college professor and elementary teacher reviewed the projects to determine which best suited the objectives of the third grade class. A WebQuest on rocks was chosen as the project to be used by the class. The WebQuest chosen included the following Introduction from the WebQuest which describes the project and the students' task:

> Mr. Johnson wants to build his house out of rock so that it will be sturdy and last forever. Mr. Johnson and his wife, Mrs. Johnson, live in California. He has contacted three builders in his area that are willing to take on the job of building his house. The first builder is Mr. Sandman, who received his degree in architecture from the University of San Diego. Mr. Sandman suggests that Mr. Johnson use sedimentary rock to build his house. The next builder who has been contacted by Mr. Johnson is Mr. Vol Cano who received his degree from the University of Hawaii and suggests that Mr. Johnson use igneous

> rock for the construction of his new home. The third and final builder that Mr. Johnson contacted was Mr. Slate, who graduated from the University of Colorado with a degree in Architecture and a minor in Geology. Mr. Slate, true to his name, recommended the use of slate. Each builder is in favor of their specific building material and is trying to get the contract to build Mr. Johnson's house. They must each try to convince him that their rock type is the best and then choose a specific rock to use, but there's a catch! Mrs. Johnson is not too excited about her husband's ideas for their new house and is trying to convince him that they should not build their house out of rocks at all (Introduction of WebQuest created by Katie Chumas).

The professor and teacher reviewed the WebQuest to be used prior to students having access. First, minor editing was undertaken and grammatical errors were corrected. Second, since the student created the WebQuest for a hypothetical situation, the professor and teacher removed information that had no relationship to the real WebQuest such as "can't use your textbook," since the students would not use a book with this WebQuest. Third, some items were reworded so that third graders could understand them such as "eg." was changed to "for example" and some sentences were shortened. Fourth, a link was added to a website one of the other students found called Rock Hounds. Fifth, sites used by Mrs. Johnson were reordered so that the sites with the most useful information were first. Lastly, links to the search engines such as Google and Vivisimo, were removed from the resource list to prevent students accidentally accessing inappropriate web sites.

The third grade students took a field trip to the college to participate in the WebQuest with the candidates. The candidates met the third graders in the education computer lab during their regular 50 minute class time. The third graders sat at the computers while the candidates stood nearby. There was nearly a one-to-one ratio of candidates to third graders. The third graders were given preliminary instruction from the college professor and divided into their groups. Their first task was to decide who would play each role in the group. At that point the groups went to work. Each group worked through the WebQuest getting information about rock formations. Students were provided paper and pencils for taking notes. Each group decided which material to use to build their house. One member from each group presented their decision to the class.

Two Surveys

The professor and the third grade teacher developed two surveys. One survey was for the candidates to determine their perspective in creating the WebQuests and how well the WebQuest worked. The second surveyed the elementary students on the learning they were involved with and any difficulties they may have had as a result of the WebQuest experience. Twenty candidates served as the teachers developing the WebQuests. Twenty-four third graders served as the students who performed the WebQuests. Both groups completed their surveys after the project.

The survey administered to the candidates consisted of five open ended questions. These questions were administered using the World Wide Web and an online data base. The teacher candidates completed these questions after the visit by the third grade students.

Results of the Candidate Survey

Seventeen candidates completed a survey after their experience with the third grade students. The questions included in the survey and summaries of the candidates' responses are reported in the following sections of this chapter.

What did you find to be the most positive thing about the Web Quest?

The candidates that responded to this question actually interpreted it from two perspectives, from the positive aspects for the third grade students and from the perspective of the candidates as authors of their own WebQuests.

Five of the seventeen candidates commented that the third grade students had a positive learning process. It was recognized that there was both an individual and group learning process going on. The value of the WebQuest process in learning was expressed. One candidate pointed out that the third grade students had control of the learning taking place. Even other non-related lessons in social skills development occurred.

As one candidate commented, "If created correctly, WebQuest will undoubtedly foster a thought provoking challenge for the students. Furthermore, students seem to gain a true insight into the subject the WebQuest encompasses." The integration of technology and curriculum that takes place in a WebQuest was mentioned by four of the candidates. Research and the Internet were specific advantages reported. Only one candidate, however, commented the actual topic in the WebQuest that was chosen for this collaborative experience, namely Rocks. That candidate specifically mentioned that the students were learning facts about rocks "such as where they were formed, what materials they were made of." One last comment made by two candidates was that the third grade students enjoyed the lesson.

Regarding the candidates' concern over authoring WebQuests, their responses covered two concepts. Three of the candidates commented about the basic knowledge they attained in using the Internet for their research to create the WebQuest and compiling that information for the presentation to the third grade students. One candidate specifically commented on learning HTML code, although it was not taught in this class. The other concept mentioned was the interaction between the candidates and the third grade students. Since this class is early in their educational career, the candidates have not had any field experience yet. Working with young students was a foreign situation for them. At the beginning of the class, they were anxiously lined up in the back of the room waiting to find out what to do.

What did you find to be the most negative thing about the Web Quest?

The candidates related this question to their own experiences. The most negative thing (cited by six candidates) was the amount of time it took to create the WebQuest, specifically doing the research to find appropriate web sites and creating and assembling their web pages. As one candidate said, "It is worth it but I am not sure that a teacher has the time to form it or the computers in the classroom to use it!"

The next most common response (three candidates) related to the quality of the resources that were found. Candidates addressed the lack of appropriate grade level information and lack of information overall in creating their own WebQuest. One candidate commented that information available in the WebQuest that was conducted was not adequate and that "we had to come up with our own knowledge, based on our own knowledge."

The less common comments revolved around specifics in creating the web pages. Two candidates did not like using Netscape Composer to create web pages. Another did not like the topics that were chosen. One candidate stated that this was a difficult project but "maybe I didn't pay enough attention to the instructions, or we didn't spend enough time on them, but it was quite difficult. But the end result, working with the kids, was worth it."

What did you learn as a result of having the students use our class Web Quest projects?

The candidates gained a variety of insights into incorporating WebQuests into instruction as a result of this project. Foremost, they gained experience "road testing" the WebQuest with real students. Within these six comments, recognized issues were that one-on-one instruction or guidance may be necessary, the steps should be clearly stated, fine-tuning might be necessary and that it is important to make the "subject flow."

The next set of comments (by four candidates) related to how the candidate learned the value of WebQuests. The way that WebQuests integrate technology and the curriculum into a real situation made this a great way to teach. WebQuests were seen as a valuable way to teach. Interestingly, one candidate commented that WebQuests can be useful for teaching small children as if this were the only age group for which WebQuests would be appropriate even though the candidates were introduced to WebQuests with a lesson prepared for the candidates in their own class to demonstrate what a WebQuest was.

The candidates also learned that they need to be concerned with the quality of the web pages that are used as resources for the WebQuests. The candidates became aware of and concerned over readability of these web pages when they found out that the students had difficulty reading some of them.

Two candidates commented that the value of doing this project allowed them to see how it would work with real students. Just knowing that WebQuests exist and that they are qualified to create them was not enough. The reality check of actually working with third grade students made WebQuests a real teaching tool for them.

Again with this question, the topic that was taught (rocks) was not that significant. Only one candidate mentioned the topic explicitly. The value was in the enjoyment that students experienced and the learning, both by the third graders and the candidates.

What did the third graders have the most difficulty with?

The most common response regarding difficulties for the third graders was based on the reading level of the web sites used in the WebQuest. Eight of the candidates mentioned reading level specifically. In fact, the reading level for the WebQuest resources were at or above the eighth grade level based on a tool built into Microsoft Word that calculates Flesch-Kincaid Readability Statistics. The web pages created by the candidate used in this project were from the 6.9 to 10.5 grade level.

The second most common comment related to the questions that the students were to answer. Four of the candidates recognized that the third graders understood the questions that were being asked on their worksheets but had a hard time finding the answers in the web page resources.

The last comments registered by the candidates were less common. Two candidates mentioned attention span and staying on the topic. The last comment was that a candidate did not know what was expected of him/her in creating the WebQuest.

What would you suggest should be changed for your next Web Quest?

When asked what they would change, the candidates broke this down into two categories: 1) those things that affected the third graders, and 2) those related to the candidates and their creation of the WebQuest project. Two candidates said that nothing should be changed.

The most popular comment (four candidates) regarding the WebQuest project that the candidates created was that there should have been more of a choice in the topic or no topic specified. They felt that their creativity was limited and that they did not really have ownership in the WebQuest. One candidate wished that the class had taken one day to practice the WebQuest. Another candidate wished that more in-class time had been allowed in the actual creation of the WebQuest. One candidate felt that more time should be spent teaching how to make the WebQuest.

The most frequently suggested change with regard to the third graders (three candidates) was using web pages with a more appropriate reading level was essential. Roles in WebQuests should not require a specific gender if possible. One male third grade student was upset because he had to be Ms. Sandman, he did not want to be a female. Another candidate felt that there should have been more purpose to the WebQuest. "Make it a little clearer, or I guess insist that we have a specific goal to the WebQuest, rather than just having kids look for answers to questions. I, and I think many others in the class, did not pose a judgment question as the ultimate goal of the project."

Other Specifically Noted Comments

- I think that it is a great project to use if you have the time and the resources.
- It was a fun project and a good resource.
- I really liked this project overall, it is a great learning tool.
- Overall, WebQuest was a great introduction to a tool that has great possibilities in the future of teaching.
- This was a great experience. Web Quest is a great learning tool for the classroom to enhance our students' education. Having the class view the WebQuest and watching the students learning and relating to the topic, made the effort and time worth it.

Results of the Elementary Student Survey

The third grade students completed a paper and pencil survey when they arrived back at their own school. The questions and their responses are reported in the following sections of this chapter.

Did you think the WebQuest was easy?

With this question, there was a split between the students, thirteen saying "yes" and nine saying "no." Even though this question did not ask why, two of the students that chose No volunteered a reason. One student commented on the hard words and the other stated that he/she could not find the answer to question 3.

What is the most important thing you learned?

The primary answer to this question revolved around rocks. Seven students commented that they learned about rocks but nothing specific was mentioned. Six of the students were more specific mentioning how rocks were formed. Four students said the most important thing they learned was how rocks can be used. One student recognized that not all people like rocks (Mrs. Johnson). One student mentioned cooperation, seeing things (information?) and wood. The last two answers were unusual: "The quiz is the most" and "It has a lot of games."

What gave you trouble in this lesson?

Seven students said they had a hard time finding information, and more specifically, about their rock. Four students claimed that reading the information was the most problematic aspect of the WebQuest. Four of the students specified a particular question that gave them trouble (mostly number four). Two students said that they did not have any trouble. Two students mentioned the quiz (a web site called Rock Hound) even though we did not give one to them. One student could not learn about wood (a possible material that could be used to build houses) instead of rocks. A student did not learn where rocks come from and the last stated that he/she "had a little bit of trouble" but did not specify why.

Would you like to do this kind of lesson again?

Students overwhelmingly answered "yes" to this question. Seven students specifically said it was fun. Seven students claimed to like this method because of what they learned. Four students just said they liked learning this way with out giving a reason. One student qualified his/her answer by saying yes as long as they learned something new. Only one student said that he/she would not like to do this kind of lesson again because it was hard.

How would you make this lesson better?

The most popular answer for this question with ten students responding was more time. Two students specifically stated making the lesson easier would make it better. Three students said the lesson would be improved if it were on a different topic (longitude and latitude, math or more choices). Two students wanted a different reading level so children could understand. One student wanted more directions on how to do the lesson. One student did not like the group work and wanted it to be individualized. Two answers were less useful ("By looking for your rock and printing it" and "By getting more games" from a web site called Rock Hounds).

Utilizing WebQuests, what else would you like to learn?

When asked specifically what would they would like to learn, a variety of answers were reported:

- I would like to learn about longitude and latitude
- About China
- About space
- How waves are made
- I would like to learn about different languages
- I would like to learn about the sunken continent of Atlantis
- I would like to learn about other rocks
- I would like to learn about tigers
- I would like to learn about owls

Did you enjoy your trip to the College?

All students answered this question in the affirmative. Student responses to this question mostly favored the fact that the students thought the field trip was fun:

- Yes because it was cool to use those computers
- Yes, I sure did. It was fun and cool
- Yes, but I want to go to different classes
- Yes because I got to see the College

- Yes, because I learned a lot of things like about a lot of different rocks
- Yes, because I liked meeting the students and do a WebQuest
- Yes because the computer is awesome
- Yes because I would like to play on the Internet a little bit
- Yes because you get to work with teacher candidates and we worked on computers

Conclusion

Overall, this project was a success, but not without problems. Both the third graders and the candidates learned as a result of the project as did the authors. Problems can be summed under a few headings. First, the college computers do not have any filtering software installed in them. One fear expressed was that the students could accidentally access controversial websites using search engines, although none did. However, since the children were supervised one-on-one by the teacher candidates, it would not be likely that purposeful accessing of the wrong web sites would take place, but search engines might have turned up something inappropriate.

Second, the readability level of the pages that were used was a problem that was obvious to both the third grade students and the candidates. Some of the terminology and scientific names caused part of the problem, but finding useful web pages at an appropriate reading level is a must. Candidates were taught how to copy/paste web page text into Microsoft Word to check readability but did not use that knowledge or ignored the results.

Third, time or lack of it, became a problem. This project was conducted in one fifty-minute college class, but should have spanned more than one session. Unfortunately, time constraints of the lab and the candidates' schedules did not allow a longer time period that day and the cost of setting up another field trip prohibited a second visit by the third graders. Having a professional development school near the college campus would be an ideal situation.

Fourth, these future teachers were unaware of what topics third grade students were studying. Although the candidates had access to the state standards, they lacked the knowledge necessary in understanding how children learn. The candidates also had a difficult time writing web pages on a third grade level. Many of their sentences were too long and complex for the students. Sentence structure needed to be simplified, while maintaining the necessary content and terminology of the lesson. They had difficulty in finding appropriate pages at a proper reading level and even creating their own web pages.

From the standpoint of the third graders, there was limited success. The students did learn some information about their specific rock type, but not about all rock types. They did, however, enjoy the experience. As one third grade student said "This was the best field trip I have ever been on!"

This project served as a pilot for developing this process more fully. Much was gained from this experience. WebQuests are worth doing, but with more preparation and attention to details such as readability level, pre meeting preparation, and follow up activities. The candidates got a reality check on some of the problems that they need to be aware of. Hopefully, after this experience, they will be more willing to integrate technology into their own curriculum with their own students.

References

Brooks, D. W., Nolan, D. E., & Gallagher, S. M. (2001). *Web-teaching: A guide to designing interactive teaching for the World Wide Web*. New York, NY: Kluwer Academic/Plenum Publishers.

DeVincentis, C. & King, P. A. (2000). Election 2000 and beyond. *Learning and Leading with Technology*, *28*(2).

Dodge, B & March, T. (1997). *Some thoughts about WebQuests*. Last retrieved September 17, 2004 from http://edweb.sdsu.edu/courses/edtec596/about_WebQuests.html.

Dutt-Doner, K., Wilmer, M., Stevens, C., & Hartmann. (2000). Actively engaging learners in interdisciplinary curriculum through the integration of technology. *Computer in the Schools*, *16*(3/4), 151-166.

Jonassen, D. H., Peck, K.L., & Wilson, B.G. (1999). *Learning with technology: A constructivist perspective*. Upper Saddle River, NJ: Prentice Hall.

Lin, B. & Hsieh, C. (2001). Web-based teaching and learner control: A research review. *Computers and Education*, *37*, 377-386.

Summerville, J. (2000). WebQuests. *TechTrends*, *44*(2), 31-35.

Tiene, D. & Ingram, A. (2001). *Exploring current issues in educational technology*. Boston, MA: McGraw-Hill Higher Education.

Yoder, M. B. (1999). The student WebQuest: A productive and thought-provoking use of the Internet. *Learning and Leading with Technology, 26*(7), 6-9

Integrating Technology into a Two Continent Consortium: Lessons Learned Teaching the TIRES Virtual Seminar

Sarah J. Mahler
Florida International University

Abstract:

This chapter discusses the challenges faced and overcome when a group of faculty needed to design and implement a "virtual seminar" that would be taught by professors at eight campuses on two continents. While the content of the course–international migration in comparative perspective between the U.S. and Europe–was important to teach, the designers were particularly concerned about building into the seminar opportunities for students to interact across campuses. The rationale was that through multiple types and times for interaction students and faculty would be exposed to an array of perspectives and their learning experience enhanced. The designers turned to several technological tools, namely, electronic bulletin boards, email "pen pals," and moderated on-line chats, to promote interaction. The chapter details both the successes and stumbling blocks encountered in the development and implementation of this seminar.

Introduction

It was the first week of August and around the table sat a dozen people–faculty, technology support technicians and administrative personnel. Their mission? Design and implement a course to be taught at eight campuses on two continents with up to ten-hour time differences and for students whose language dominance ranged from English to French, German, Dutch and Spanish. When would this course begin? In less than three weeks! Mission Impossible? Almost–if not for the innovative use of technology. This chapter discusses what this group of people accomplished in order to successfully teach a "virtual seminar" coordinated among all the campuses.

The Mission

Florida International University (FIU) won a grant from the U.S. Department of Education, Fund for the Improvement of Secondary Education (FIPSE) to serve as lead partner of a federally funded consortium linking four campuses in the U.S. with four campuses in Europe, the latter funded by a matching European Union grant. The consortium's focus is studying immigration and xenophobia in comparative perspective on the two continents. Its acronym, TIRES, stands for "Transnationalism, International Migration, Race, Ethnicity and the State." The easternmost member and lead partner for the European side is a German university on the Polish border and the westernmost partner is the University of Colorado in Boulder. The TIRES consortium, with some alternations in configuration, had functioned for some half-dozen years prior to receiving the funding and primarily facilitated undergraduate student exchanges. The grant support stipulated that the consortium partners would execute three elements: (1) undergraduate student exchanges aided by grants to help offset students' added expenditures for travel and/or to acquire language proficiency; (2) creation of three-week residential summer institutes for graduate students; and (3) the development of a "Virtual Seminar," a course that would be taught at each campus by TIRES faculty.

As conceptualized in the grant, the Virtual Seminar (VS) would combine a face-to-face course taught by local faculty at each campus and computer-mediated communication among students and faculty across the consortium, but particularly among those involved in VS courses taught during the same semester. Thus, the VS would not depend entirely upon an individual faculty member (or in some campuses multiple TIRES faculty members) to teach all the course material as in a standard course. Rather, the VS would capitalize on different TIRES faculty expertise. In other words, the course design would bring TIRES faculty with specific knowledge to "teach" this part of the seminar and the local faculty member would then teach only his/her own areas of expertise while the other TIRES faculty would fill in. For example, if a faculty member's expertise were in theories of migration, she would handle this section of the course, and if another member's expertise were in xenophobic reactions to immigrants, he would "teach" this section. The critical problem facing the VS planners sitting around the table in early August was *how* to operationalize this collaborative plan. Moreover, the VS part of the grant proposal emphasized that the VS would provide opportunities for interactions between students at the different campuses and particularly EU versus U.S. students, and interactions between students and faculty across the consortium. This promise accentuated the difficulty of the task facing the VS planners. What would they design?

Scholarly Precedents

The challenges faced by the VS planners are both similar to and different from those faced by other educators pioneering computer-mediated learning environments. A key similarity is the desire to build student-student and student-faculty interactivity into on-line and web-assisted courses (Cafolla & Knee, 1999; Davidson-Shrivers, Tanner, & Muilenburg, 2000; Freiermuth, 2002; Irvine, 2000; Jeong, 1996; Schwier & Balbar, 2002; and Wang & Newlin, 2001). Roberson and Klotz (2001), for example, are concerned that:

> often on-line courses are configured and delivered in a style more often associated with independent study or correspondence work, i.e., students working independently to complete posted assignments at their own pace. While this format may work in some instances, it leaves a 'missing link' in the learning curve for students since they lack the opportunity to benefit from the experience of structured dialogue and sense of community that can be created in a traditional on-site classroom environment (p. 1).

Courses delivered completely on-line are identified as those most in need of filling in the "missing link" of interactivity. Synchronous and asynchronous technologies have been introduced to meet this need such as email correspondence including listservs, electronic bulletin boards, video conferencing and chat rooms. Though the research on the effectiveness of these different technologies is still in its infancy, there is consistent evidence for a basic conclusion, viz., that interactivity improves student satisfaction particularly students' sense of a learning "community" which otherwise is largely absent (Orey, Koenecke, & Crozier, 2003; Roberson & Klotz, 2001; Schwier & Balbar, 2002). This may ameliorate the current high drop out rates of students in on-line courses (Hill, Han, & Raven, 2001). Additionally, computer-mediated communication (CMC) in courses facilitates higher participation rates by students overall, since the relative anonymity of the on-line environment helps shy students overcome their inhibitions to participate, inhibitions most evident in traditional face-to-face learning environments (Irvine, 2000). Research that compared different technologies for their promotion of learning suggests that listserv discussions and/or threaded electronic bulletin board discussions promote greater student reflection and quality of interaction (Davidson-Shrivers et al., 2000; Dietz-Uhler & Bishop-Clark, 2001; Irvine, 2000; Jeong, 1996; Roberson & Klotz, 2001). With the exception of Dietz-Uhler & Bishop-Clark (2001), the same studies typically found the chat format to be more dynamic but prone to student digression from the discussion topics.

The TIRES faculty planning the VS did not enjoy knowledge of the literature on CMC. Our mission to design the VS plunged us into this field and, in retrospect, we were able to achieve our goal of consortium-wide interactivity by employing different technologies, some in ways that other scholars in the field do not appear to have pursued. As described in detail below, we promoted student-student and student-faculty interaction via email (through pen pal assignments), threaded bulletin board discussions, taped lectures (not video conferencing but which at least introduced students to TIRES faculty), and, most importantly, *moderated* chats. Moreover, we integrated these technologies not into on-line courses, but into courses that were taught in brick and mortar classrooms. This makes the VS endeavor distinct from most of the literature on CMC which addresses issues specific to on-line teaching environments. After I discuss our experience in detail, I will return to the issue of moderating chat rooms, as anticipated immediately above.

Background Continued: CMC Precursors to the VS

In the spring semester of the year before the VS was to be implemented, faculty at two TIRES campuses (FIU for the U.S. and the University of Liege [Belgium] for the EU) attempted to provide interactive experiences for their students studying international migration. This experiment helped identify both possible modes for interaction during the VS and also some pitfalls to avoid. The FIU course was web-assisted through the software package WebCT (similar to Blackboard). FIU students would log into WebCT where they could access course material, bulletin boards and chat rooms. The bulletin boards in WebCT are threaded and thus can hold iterative discussions. They have an added advantage–they can be edited and manipulated by the instructor ("designer" in WebCT lingo). For example, students' postings that are off-topic can be deleted and postings that are improperly placed (e.g. posted to the wrong discussion thread) can be moved. The chat tool in WebCT, however, is not sophisticated. It is just a chat room that anyone who has logged into the WebCT course can enter. That is, it is an unregulated or unmoderated room.

The WebCT format worked fine for the FIU students and their TIRES faculty instructor. However, it was problematic for interactions with the Liege students. Why? This has to do with the log-in security features of WebCT. At FIU and probably on most campuses, students must be registered in a course at the university in order to have access to the WebCT components of the course. At FIU, this is an automated process where students' FIU email addresses become their WebCT log-in usernames and their student IDs become their initial passwords. This became a stumbling block for the Liege students who were registered for their own class in Belgium but not for the FIU course. The WebCT staff at FIU had to be contacted and each Liege student given a WebCT username and password in order to gain access to the bulletin board. Liege students then were asked to post their ideas to the same bulletin board within WebCT as the FIU students. This worked well once the log-in process was solved. However, the quality of the interaction was low because it was not in real time and students could not get to know one another since they could only respond to intellectual issues.

The Virtual Seminar Course Model

On that early August day when the emergency implementation team met, expectations for more satisfying levels of student-student and student-faculty interactions punctuated the discussion about how to design the VS course. Moreover, the course needed to maximize flexibility, permitting faculty to teach it according to their academic calendar (calendars for each campus varied widely) and their own interests and didactic approaches. The basic format that was sketched out by the planners and then implemented was the following:

(1) The course would be taught in English, the common TIRES language.

(2) Professors would develop course modules that would
  a) feature their own expertise and
  b) complement each other so that there would be minimal overlap.

The modules would be self-standing, meaning that they could be taught independently of each other and thus utilized in any order. Faculty could utilize several or all of them as building blocks for their own VS course versions. Finally, faculty teaching the VS would retain full control of the course content and the grading protocol because they would have to issue grades. In sum, the VS format was designed to be exceedingly flexible for teaching and grading purposes but with a common content and shared vision for technology-aided interactions.

Individual Modules

Planners decided that each module would include several standard elements. These included a (1) mini syllabus, (2) videotaped lecture, and (3) interactive components. The mini syllabus included a description of the module, a list of required readings, an outline of the lecture, and a list of discussion questions students would be expected to address in the interactive component section. The readings were made available either in hard copy (sent to the lead university on each continent and then distributed) or scanned and made available in PDF format on line. Lectures were videotaped in most cases at FIU when the TIRES faculty attended the summer institute for graduate students. They were made available to consortium members in the format they desired: DVD, CD, VHS or PAL. Each module author could also suggest some assignments that students could do, including interactive assignments. These module elements were assembled into a specific module webpage accessible from the TIRES VS homepage. Technical support staff at FIU designed an applet log-in function that would pop up whenever anyone clicked the "VS" icon on the main TIRES webpage. Once through the log-in process, students would have access to all the VS modules. Only enrolled students would be given the correct log-in information to help keep the VS protected from uninvited visitors.

Thus, when students' local VS syllabus called for them to use a certain module, they were directed to the VS webpage and told to select the appropriate module. By clicking on the module icon, they would come to its webpage where they would find icons for the mini syllabus, a bulletin board and a chat tool. Students would do the required readings and address the discussion questions posed on the mini syllabus, unless otherwise directed by their own VS course instructor. Additionally, students would watch the videotaped lecture for that module. The latter helped students make a personal connection to the module author and prepare them for interactions through the bulletin board and chat tool on the module webpage. Given that these lectures were taped, they could be viewed again and on demand. Additionally, though this was not implemented, they could be made available digitally so that students could access them through the VS webpage or through WebCT.

Use of Technology: Interactive Components of Modules

Though the VS employed technology through the web pages already mentioned, its design called for more sophisticated uses that I will discuss now. Building on the previous FIU-Liege experience of the bulletin board interactions, each VS module incorporated a bulletin board that could be used by all classes using that module. Moreover, the planners heard about the advantages of *moderated* chat rooms from a couple of colleagues at FIU. They discussed this at the August meeting and then decided to implement the use of a moderated chat tool for real-time student-faculty interactions–chats on the content of each module that would link the module author to the students studying it. I will now turn to the conceptualization, operationalization and real experience we had with each of the interactive technologies.

Module Bulletin Boards

As indicated above, each module author developed a series of discussion questions and included them in his or her mini syllabus. The expectation was that these would become the basis for bulletin board threaded discussions. Since the VS was taught on different campuses and according to different school calendars, these discussions were expected to be asynchronous but still focused. That is, one set of students studying a particular module would always be the first. They would encounter the discussion questions posed by the module author inside the otherwise empty bulletin board for that module and begin posting their responses. At some later, usually unknown, time another class would use the module. This second class would encounter the responses already posted and be able to react not only to the discussion questions posed but also to the responses by the previous set of students. Over time, more and more VS classes would encounter the bulletin board and would presumably benefit from the expected variation in viewpoints given students' high diversity of national backgrounds and local circumstances. At FIU alone, most students are either themselves immigrants–primarily from Latin America and the Caribbean–or the children of immigrants.

The modules' bulletin boards proved to have advantages and disadvantages. One of the advantages is anticipated in the discussion immediately above, namely, that students' postings would accumulate within bulletin boards over time and thus enrich the discussion. Even students who studied the module early could return at any time and revisit the discussion if so desired. Second, the bulletin board postings have the advantage of creating a permanent record of students' performance, facilitating the grading of this performance.

However, we also found that the bulletin boards suffered from several disadvantages, most of which are attributable to the bulletin board software. First of all, the bulletin board tool that FIU technical staff copied onto each module was very basic. Unlike the bulletin board in WebCT, instructors could not modify it given its location on web pages that could not be modified by instructors. While this restriction made sense given the fact that no one VS class faculty instructor "owned" the bulletin board, it did mean that discussions proceeded unedited and thus on occasion became off topic. Secondly and related, any problems or issues arising from the bulletin boards had to be solved by technical staff who required payment for their efforts. The grant provided some funding for this assistance, but it was largely depleted by paying for the design of the VS and the development of individual web pages. Third, instructions for how to use the bulletin board (and also the chat room) were written by FIU TIRES faculty to save money and were limited. The instructions as well as the VS homepage included links to "technical assistance" but these links did not lead to real-time help and the requests were only forwarded to a TIRES administrative aid at FIU who then had to seek further assistance. This meant that much time could elapse before users could get the help they needed, particularly if they accessed the module during non-business hours in Miami–very likely given the geographical distribution of the consortium.

Additionally and critically, the bulletin board suffered from the fact that student postings recorded their name but not university affiliation. This hampered the efforts of other students to look at what, for example, the German students thought on an issue versus the Belgians, French or Americans. This particular problem could have been avoided easily if students were instructed to log in not only by name but also by institutional affiliation. Finally, the bulletin boards proved less useful than anticipated because of the lack of TIRES faculty coordination. That is, no one faculty member was assigned the task of ensuring that each module's bulletin board was loaded with the correct discussion questions. Instead, either the instructor checked to make sure that they were in place prior to assigning the module or, in the worst case scenario, assumed they were in place and only found out to the contrary if students reported them missing.

Moderated Chat Room

The most successful–but also the most difficult–technology integrated into the VS was a moderated chat room. When the VS planners met in August, the TIRES faculty were unaware that there are commercially available moderated chat tools. We were told about one in development by two FIU faculty members, however, and immediately saw the possibilities of incorporating this type of technology into the VS. A moderated chat room is one in which someone–presumably an instructor or, in non-academic circumstances, an overseer–keeps watch over the chatting activity and keeps it either on track or away from prohibited topics. In the VS case, faculty were concerned that in past experiences with students in chat rooms they would often begin on topic but drift away quickly and that there was no way in an unmoderated chat environment to limit or avoid such tangents. In the chat room technology designed and developed by two FIU faculty members[1], students could log into a chat room and find a public chat area.

The difference between a moderated and an unmoderated chat room is the fact that the moderator receives each message and decides if it should be permitted into the public chat area. In the case of TIRES, the same faculty member who wrote the module that the students were studying typically also moderated the chat room environment. The author/moderator would receive each student's posting, almost invariably a response to a specific discussion question from the mini syllabus or from the author posting a different question to the public chat room, and decide if it were pertinent to the public chat room's ongoing discussion. The moderator would receive all postings in a special inbox viewable only by the moderator. S/he would click on a posting that would then jump from the large white "inbox" on the moderator's panel into the "Pending Chat Text" window. She then would click on one of the boxes to the right of the white inbox for how to handle the posting: (1) accept it which would send it to the public chat area, (2) reject it with a note to the student as to why it was rejected which would send the note only to the student, (3) ignore it which would send it to the bottom of the list of postings in the "inbox," or (4) delete it without letting the student know why. The designers of the moderated chat software included in the tool a group of pre-determined moderator responses that the moderator could send back to students to explain their posting's rejection, such as "stay on topic," "rephrase more clearly," and so on. These options can be seen in the top right area of the moderator's panel. The moderator's screen includes both the student's view (or public) panel and the moderator's panel.

If the moderator sends a message to an individual student, s/he would receive this message in his or her view, just below the window holding the public chat room. Because the moderated chat tool only permits accepted postings into the public chat domain, students send messages and do not know why they do not show up in that public area unless they receive a response from the moderator. The moderator, thus, not only controls the quality of the content inside the public chat area, but can also provide feedback to students.

The TIRES experience with this moderated chat tool proved three things. First students and faculty really enjoyed the chats and they found them to be much more stimulating than the threaded bulletin board discussions. Second, in order to use this technology one had better make sure that it will not fail at the wrong moment. Indeed, we had several failures during chats that plagued the TIRES experience. These were due to equipment error. In one case, the server running the chat tool crashed making the tool mysteriously unavailable, and on several other occasions the software failed during the chat, apparently reaching its maximum load factor when more than a dozen participants were logged in. In the latter case, some TIRES faculty knew how to restart the tool but this was too technical to teach all the faculty.

Third, moderated chats are highly dependent upon the typing, concentration and language skills of the moderator. Moderators would receive into their "inbox" the students' postings, often many at a time. Each posting requires reading it, deciding what to do with it and implementing that decision. Furthermore, moderators have to keep track of the general discussion going on in the public chat area in order to know what to decide to do about each student's posting. These decisions are cognitively stressful, particularly if the moderator's first language is not English. Moreover, moderators need to be good typists in order to referee the discussion, keeping it on topic and switching topics. The structure of the moderator's panel illustrates that the instructor must shift his/her hands from operating the keyboard to using the mouse. This back and forth activity is constant.

To give a concrete example, imagine there are 15 postings sitting in the moderator's inbox. She scans them quickly and decides which one is most appropriate to continue the discussion in the public chat area. She clicks on this posting and selects the "accept" button with her mouse. It then heads to the public area. She then clicks on several other postings, one at a time, that are not well phrased and clicks the "delete" button. She then decides that she should send several other responses back to students with comments; each time she does so, she clicks on a posting from her inbox and then selects "reject with comment" and then types a comment in the box labeled "Text to Send" and hits return. While she is doing this, numerous other messages arrive in the inbox. She ignores them for the time being, more concerned about shifting the discussion to another topic. She types this topic into the Instructor's "Type/Select Message" area box and hits "return" so that it appears in the public area. She then turns to dealing with the postings that have accumulated. The constant pressure to process postings is exhausting. I have moderated several chats lasting over an hour and up to two hours. The time flew by but I was completely incapable of doing anything afterward! Fortunately,

I am a fast typist and very knowledgeable about the chat tool itself, having helped to implement a series of improvements after participating in a few chats with a beta version of the chat tool. Imagine how much more difficult it was for colleagues for whom English is their second language and whose keyboards are not optimally configured for typing in English.

Advantages & Disadvantages

Aside from these major lessons learned, there are a host of other advantages and disadvantages we experienced with the moderated chat tool. I will begin on the positive side. Among the greatest rewards was student response that was overwhelmingly enthusiastic. The only complaints we heard were that sometimes there would be little activity in the public area–presumably when the moderator was busy handling the postings–and some students complained that their postings never made it to the public area. Another plus to the chat tool was the opening up of opportunities for real-time interactions. Students could "meet" their module instructors via the web in addition to via videotaped lectures. Of course, given the major time differences, chats had to be scheduled carefully and often could not occur during class time. The latter was particularly true in Europe where the universities in the consortium by and large did not enjoy computer labs that could be scheduled for classes and in which there were sufficient computers available so that all students could log on simultaneously. This hurdle was overcome in some cases by the fact that students could log into the chat room via the VS homepage from any computer with Internet access. Indeed, faculty also found themselves moderating chats from home given that they might have to handle the chats in the evenings (the European case) or early mornings (the U.S. case). Students on the U.S. side enjoyed more ready access to computers in general than the Europeans. The most affected campus was the German that has fewer than 10 computers with Internet access available for student use and no computer labs.

Certainly the chat room is a cost-effective interactive tool, much less expensive than a satellite hookup. Another advantage of the chat room over the bulletin board in the TIRES case was that students were instructed to log into the chat room using their real names. When they clicked on the chat room button from the module webpage they would have to log in and they did so with their own names. The chat log could be saved before exiting out of the program and through the log VS instructors could revisit the chat and also examine the quality of each student's postings. The log also served as a "product" that was important to provide to the funding agency to show how their grant moneys were being put to use. Finally, to reiterate points made above, the tool proved invaluable for focusing the interactive discussions, to keep them from straying away from the module's content.

Arguably the worst problem we experienced was the fact that the chat tool software failed several times–and given the difficulty of scheduling the chats they could not be made up. After the first failure, FIU TIRES faculty worked hard with technicians to provide a more stable server environment and to make sure that a technician was on call in case of a failure. The housing of the tool on an FIU server meant, however, that all chats, even those not involving FIU faculty, needed to be coordinated with FIU so that the technicians were on watch. This added strain on the U.S. lead partner but it was the only campus with readily accessible technical support staff familiar with the TIRES program and technology. An alternative for anyone interested in moderated chat rooms is to find a commercial one offering 24/7 support. This comment leads to another, albeit inevitable, drawback, namely, the fact that the complexity of the moderated chat tool requires moderators to be adequately trained prior to its implementation. In the TIRES case, a CD with instructions was prepared and sent to faculty and then they practiced moderating the tool before using it with students. This did not prove sufficient in every case.

Lastly, as discussed above, the moderator has a heavy cognitive load with this technology, particularly when the number of students participating rises above 10. As a consequence, he cannot be expected to keep track of the fact that certain students' comments are permitted into the public area more than others or that some students' postings are never accepted. A possible remedy for this situation would be to assign the students' regular instructor, not the module author, to track the discussion and to send a message to the moderator about any students who do not seem to be participating, at least as revealed in the public area window. In the TIRES case, the regular instructor can only do this while logged in to the chat room just like the students, but her message to the moderator could be written in a particular way, such as in capital letters, to communicate quickly to the moderator her concerns.

Other Technology-Assisted Interactions in the VS

When I taught the VS with my colleague, Associate Provost A. Douglas Kincaid, we employed a few additional technologies than those already described above. I will now briefly describe them and their usefulness. We taught the VS as a web-assisted course using WebCT and we took advantage of its more sophisticated bulletin boards to facilitate student exchanges around the course's topics. We did this in several ways through the bulletin boards. On many occasions, we had students post individual responses through the boards as described earlier. This produced threaded discussions. Students were informed that they would be graded for their performance in these discussions using a grading calculus as follows. We gave them separate evaluations for (1) the number of individual entries each student made into the discussion; (2) the number of separate times they logged into the discussion to participate (this information is readily available given that each posting has a time and date); and (3) the quality of their interventions. These evaluations were averaged to provide a grade for each discussion assignment. Secondly, we assigned students into discussion groups and had them focus on a topic from the perspective of a country. For example, we would be studying xenophobic reactions to immigrants in different countries and were looking for country-to-country comparisons. We thus formed groups (WebCT has a randomizing function to create groups) and assigned each group a nation to research. Through WebCT we also created a special bulletin board for each national group. Students were assigned the task of communicating their research and findings through this bulletin board on WebCT. Thus, the work could be inspected by each group member and also by other students and as well as we, the faculty. We then developed a grading protocol for these assignments as well.

Additionally, we required students to access the Internet to do much–but by no means all–of their research and encouraged them to be in touch with TIRES faculty and students in other countries to aid them when they had to find information on Europe. They also had to subscribe to listservs sending information on the course topics and to post important articles and information to topical discussion bulletin boards. We encouraged students to present their group projects using technology, primarily PowerPoint. This was aided by the fact that our class met in a specially designed classroom equipped with direct Internet access, a large broadcast screen, WiFi technology, and more audiovisual aids.

A final use of technology in this particular class involved an interactive assignment linking our FIU students to students in Liege. We obtained a class roll from Liege and assigned each of our students to a Liege student to become interactive Internet "pen pals." Twice during the course of the semester we required these pairs to conduct a discussion on a particular set of questions related to a module that each class was studying. The "pen pal" idea was good but it ultimately failed for a couple of reasons. First and foremost, it was implemented during and not prior to commencement of the semester. In the FIU case, we structured this assignment into the students' grading scheme while this did not happen on the Liege side. Consequently, the Liege students were not as consistent participants as the FIU students, the latter complaining when their emails would not receive responses. Second, differences in academic calendars were not anticipated adequately and proved problematic. During the second pen pal assignment, for example, the FIU deadline fell, unbeknownst to us, during spring break at Liege. FIU students panicked due to non-responses and the assignment had to be extended. These problems are easily remedied with greater coordination and, indeed, the following year the pen pal assignments went well. We do suggest, however, that anyone who wishes to implement such a "pen pal" idea *not* designate it with this label. Students told us later that "pen pal" evoked a very elementary school image and was not compatible with the seriousness of the assignment at the university level.

Overall Lessons Learned

The TIRES Virtual Seminar would have failed without integrating technologies into it. We chose technologies well and have, on the whole, been quite satisfied with the pathways we took. This does not mean that there is not much room for improvement. In our case, however, the improvement falls less on the side of which technologies to integrate and more on the side of adequate planning and preparation. We were very lucky, indeed, that in the space of only a few weeks we were able to put together the VS sufficiently to have it run on several campuses during the fall semester. It ran better, not surprisingly, during the spring semester. The fact that we waited so long to design and implement the VS is due to several factors, not the least of which was the necessity to bring together most of the TIRES faculty involved to ensure that whatever was designed would be amenable to them and thus implemented. We sought a democratic design process, one guided by technical expertise, and we could only accomplish this during the graduate students' summer institute that brought together the TIRES faculty. At that time one European junior TIRES faculty member was designated VS coordinator, but his location in Europe while the web pages and servers were located at FIU in Miami provide a hindrance to his ability to coordinate the VS, particularly the development of the website and the scheduling of on-line chats.

The most fundamental lesson we learned was the crucial role of preparation. We recommend advance preparation, allowing sufficient time to test the technologies and to train users of these technologies. We did not enjoy this timeframe and as a result operated largely with design leading to direct implementation–AKA "crisis mode." Even non-technological matters such as coordinating academic calendars could have been better anticipated to avoid problems experienced down the line.

We learned lessons about the value and limitations of technical support as well. FIU brought to the planning table several such individuals from its Center for On-line Training (COT). They helped steer us in the right direction but did not, themselves, know the technology as well as other programmers and systems analysts we had to consult later on when we needed to use the chat room tool. Only then did we find out, for example, that commercial products with much better support are available and might have been better for our own needs. But the COT staff members were very helpful at least in part because they could see that our TIRES VS needs was pushing the frontiers of technologically aided learning experience. As such and given their overall mission, they participated without expense to the TIRES project. However, the same was not as true for the other FIU technical staff. At FIU, as I imagine is true in most other campuses, much technological support must be purchased. A word to the wise, then, is that those seeking to integrate technology into their courses also seek to find funds to pay for it, unless, of course, faculty possess the necessary skills and knowledge.

As a final note, I wish to lift up the fact that in the TIRES experience, technology helped to unify an otherwise loosely knit consortium. In our case, such a loose configuration served well to fulfill the other

TIRES grant promises, namely, undergraduate exchanges and the graduate student summer institute. For the Virtual Seminar, however, greater coordination and cooperation was needed and we found it, aided in no small part by technology. Moreover and importantly, the technology we chose proved quite popular–particularly the on-line chats–among both students and faculty. We could really see the intellectual merit of these interactions and also enjoyed them. The proof is in the pudding: One faculty member who was very skeptical about the chats during the planning meeting in August changed her attitude 180 degrees some months later right after finishing moderating her first chat. "I'm a believer!" she wrote in an email to all the TIRES faculty.

Conclusion

Research on integrating interactive technologies into on-line instruction environments shows that "students learn better via computer-based communication than face-to-face instruction . . . [because] students must take more responsibility for, and be more active in, the learning process" (Irvine, 2000, p. 1). Our experience with the TIRES VS goes a step further to mark how Computer-Mediated Communication (CMC) can enhance traditional classroom experiences as well. Though we did not incorporate a study of student experiences or learning into our design of the VS nor did we explicitly have them evaluate the CMC components of their course, anecdotal evidence is powerful that these interactions promoted both student learning and enjoyment.

We ventured into two areas not well developed in the CMC literature that merit consideration at this final conjuncture: moderating and evaluating interactions. Though some publications to date mention the role of moderating electronic interactions (Irvine, 2000; Madjidi, Hughes, Johnson, & Cary, 1999; Roberson & Klotz, 2001), the possible value of moderating chat room discussions does not seem to be discussed to a great extent in the scholarly literature. Surprisingly, even in a study that identified a multitude of problems plaguing chats, the issue of moderation did not merit a listing (Sapone & Sing, 2001). Our TIRES experience involved moderated (VS) and non-moderated (precursor to VS) chats. Transcripts of these chats clearly reveal that the moderated chats stayed more on topic. Moderation may invoke more traditional instructor-led learning and thus detract from the greater discussion flexibility and fluidity of chat rooms (Roberson & Klotz, 2001), but our goal was predominantly to keep students focused and moderating the chat room helped us achieve that goal. In terms of evaluating student performance, TIRES faculty invented what seemed reasonable methods without assistance from the literature that is overwhelmingly silent on this issue. Some articles refer to the higher quality of student writing and thoughtfulness in threaded bulletin board discussions versus chats (e.g., Davidson-Shrivers et al., 2000; Dietz-Uhler & Bishop-Clark, 2001; Eastman & Swift, 2002; Irvine, 2000; and Roberson & Klotz, 2001), but these authors did not provide insight into specific grading rubrics for these types of interactions. After some discussion, my colleague and I at FIU developed a rubric for the electronic bulletin boards and pen pal assignments that combined frequency of interaction and quality of input.

Footnotes

[1] The chat tool is entitled "Instructor Facilitated Instructional Communication System" or IFICS.

References

Cafolla, R., & Knee, R. (1999). *Adding interactivity to web based distance learning*. Paper presented at the Society for Information Technology & Teacher Education International Conference, San Antonio, TX.

Davidson-Shrivers, G., Tanner, E., & Muilenburg, L. (2000). *Online discussion: How do students participate?* Paper presented at the Annual Meeting of the American Educational Research Association, New Orleans, LA.

Dietz-Uhler, B., & Bishop-Clark, C. (2001). The use of computer-mediated communication to enhance subsequent face-to-face discussions. *Computers in Human Behavior*, *17*, 269-283.

Eastman, J. K., & Swift, C. O. (2002). Enhancing collaborative learning: discussion boards and chat rooms as project communication tools. *Business Communication Quarterly*, *65*(3), 29-41.

Freiermuth, M. R. (2002). Internet chat: Collaborating and learning via e-conversations. *TESOL Journal*, *11*(3), 36-40.

Hill, J. R., Han, S., & Raven, A. (2001, November 8-12, 2001). *Build it and they will stay: A research-based model for creating community in web-based learning environments*. Paper presented at the National Convention of the Association for Educational Communications and Technology, Atlanta, GA.

Irvine, S. E. (2000). *What are we talking about? The impact of computer-mediated communication on student learning*. Paper presented at the Society for Information Technology & Teacher Education International Conference, San Diego, CA.

Jeong, A. (1996). The structure of group discussions in online chats. *Journal of Visual Literacy, 16*(1), 51-63.

Madjidi, F., Hughes, H. W., Johnson, R. N., & Cary, K. (1999). *Virtual learning environments* . Long Beach, CA: Pepperdine University, Graduate School of Education and Psychology.

Orey, M., Koenecke, L., & Crozier, J. (2003). Learning communities via the internet a la epic learning: You can lead the horses to water, but you cannot get them to drink. *Innovations in Education and Teaching International, 40*(3), 60-69.

Roberson, T. J., & Klotz, J. (2001). *Chat: The missing link in on-line instruction*. Paper presented at the Annual Meeting of the Mid-South Educational Research Association, Little Rock, AK.

Sapone, C. V., & Sing, D. K. (2001). *Interactive internet chat room inquiries and strategies*. Niagara, NY: Niagara University.

Schwier, R. A., & Balbar, S. (2002). The interplay of content and community in synchronous and asynchronous communication: Virtual communication in a graduate seminar. *Canadian Journal of Learning and Technology, 28*(2), 21-30.

Wang, A. Y., & Newlin, M. H. (2001). Online lectures: Benefits for the virtual classroom. *T.H.E. Journal, 29*(1), 17-18,20,22,24.

About the Authors

Teaching with Technology: Infrastructure Support and Online Discussions

Steven R. Hornik is an Assistant Professor of Management Information Systems in at the University of Central Florida. He obtained his PhD from Florida International University. His research interests includes technology mediated learning and training, social learning in online contexts, and cultural effects of TML. He has published in Journal of Information Systems, European Journal of Information Systems, IEEE Transactions on Professional Communication, and others.

Julie A.B. Cagle, Ph.D., is an associate professor in the Finance Department at Xavier University. She regularly uses technology to supplement traditional face-to-face classes and has taught two online courses in the MBA program. She serves on the Williams College of Business Teaching Development Committee responsible for teaching development programming. She also serves on the Academic Technology Committee for the University. She has participated in presentations and panel discussions on incorporating technology in the classroom and incorporating business ethics into courses.

By Instructional Design: Facilitating Effective Teaching and Learning with Technology

Donna Harp Ziegenfuss is an instructional design specialist in the Faculty Technology Center at Widener University in Chester, Pennsylvania where she works with faculty involved in integrating technology. She received her M.S. degree in Applied Technology with a specialization in Instructional Design from Chestnut Hill College in Philadelphia and is currently doctoral student in the Widener University Higher Education–Academic Leadership Program. She is an adjunct faculty member at Chestnut Hill College and she teaches in the Applied Technology graduate degree program. Donna has seven years experience working in the area of faculty development in both K-12 and higher education. Her research, publications and presentations focus on instructional design and technology-related faculty development issues.

Integrating Technology into the Curriculum

Kelly Reimer is the coordinator of the elite (Emerging Leaders in Technology @ Elon) Program at Elon University, North Carolina. Her work at Elon helps integrate technology into academics, administration, and collegiate life by employing students who mentor faculty, staff, and other students in the use of technology. She holds a B.S. degree (2000) in Business Administration and International Studies from Elon College, and she is currently pursuing her M.Ed. degree in Higher Education Administration at the University of North Carolina at Greensboro. She has presented at several regional and national conferences on the use of students for technology support services. This is her first published work.

From Talking At You to Talking With You: Reshaping PowerPoint for Interactive Learning

Ralph H. Olliges, Ph.D., possesses over twenty years experience in academia and computer technology. Currently, he is a faculty member at Webster University in Saint Louis, Missouri. He is a member of the Multidisciplinary Studies department in the School of Education. Also, he is the Coordinator of the M.A.T. Educational Technology program. For the last twenty-two years, he has taught many technology courses both face-to-face and online. His area of expertise deals with how to successfully integrate technology in the classroom. He has numerous publications and presentations with regards to adapting technology to the various student learning styles in the context of classroom teaching and learning.

Sebastian Mahfood is assistant professor of intercultural studies, director of the global vision initiative, and coordinator of instructional technology at Kenrick-Glennon Seminary. He is also a visiting professor of cyberethics at Webster University. Both of these institutions are located in St. Louis, Missouri. Sebastian has been actively engaged in the development of transactional learning methods since discovering the pedagogical value of the Internet in 1998. Currently, he is writing his dissertation in postcolonial theory at St. Louis University.

Dr. Roy Tamashiro is Professor of Education and the Director of the School of Education's Online Programs at Webster University (St. Louis, MO). He is the recipient of the prestigious William T. Kemper Excellence in Teaching Award (2000), and the Missouri Governor's Award for Outstanding Teaching (2000). He works extensively with schools in the St. Louis region, throughout the U.S.A., Europe and Asia on video and technology curriculum integration, teacher networking and professional development. His numerous publications and keynote speaking engagements have also reached educator audiences throughout the world.

Integrating Assistive Technology into Higher Education Settings

Tom Lough started a newsletter in the 1980s for teachers using the Logo computer language that grew into Logo Exchange, a professional journal published by the International Society for Technology in Education (ISTE). He coauthored the first instructional technology chapter to be included in Classroom Teaching Skills, a leading teacher textbook. He was the lead author of a writing team that produced six Logo books and served on the advisory committee for Teaching and Computers, an instructional technology periodical. He was a member of the steering committee for three international Logo conferences hosted by the Massachusetts Institute of Technology, and was the keynote speaker for a number of other instructional technology conferences. In the 1990s, Lough hosted a number of educational robotics activities at the National Educational Computing Conference (NECC). Since coming to Murray State University associate professor Lough has developed a graduate course in instructional technology with emphases on international collaboration and the incorporation of assistive technology into mainline classroom teaching. He has also developed a graduate-level teaching endorsement program in instructional technology for MSU. He hosts a series of summer robotics camps for upper elementary and middle school students. He is the author of Learning with MicroWorlds Pro.

Dr. Eric Umstead is an assistant professor of Special Education in the College of Education at Murray State University in Murray, Kentucky.

Integrating Multimedia Technology and Interactive Video into Counseling Skill Training

Leping Liu, PhD, is Associate Professor in the Department of Counseling and Educational Psychology at the University of Nevada, Reno. She teaches courses on the application of technology in education. Her research interests focus on technology integration, multimedia production, online learning environment, technology and mathematical thinking, technology and counseling skill training, and design and modeling. She has authored and coauthored eight books, established a website for a statistical textbook, published many articles in refereed journal and proceedings, and presented numerous professional papers at regional, national, and international conferences. She has been on editorial boards of three international journals, and served as a member on several national and international academic associations and institutional committees.

Dr. Paul C. Abney, a Licensed Professional Counselor and Nationally Certified Counselor, is an Assistant Professor of Education at the University of the Virgin Islands. He received his Ph.D. in Counselor Education from the University of North Texas. He currently teaches graduate and undergraduate students in the areas of counseling and guidance. His research interests include the development and application of technology programs for the training and instruction of entry-level counselors, and the use of dialogical pedagogical applications for working with diverse student bodies. He resides with his wife and daughter on the island of St. Croix in the U.S. Virgin Islands.

Teaching Disaster Management: An IT-Enhanced Hybrid Problem-Based Learning and Lecture Model

Dr Adrian Bromage (a.bromage@coventry.ac.uk) is presently based at Coventry University Centre for Higher Education Development. A psychologist by background, he joined the University in 2002, and in 2003 completed his PhD, which investigates the role of academics' prior beliefs and ongoing experiences on their perceptions of technology-led educational change, at the University of Birmingham. He is currently involved in Coventry University's evaluation of its 'Learn Online' initiative and the Enhancing Teaching-Learning Environments in Undergraduate Courses Project (http://www.ed.ac.uk/etl), which is funded by the Teaching and Learning Research Programme of the Economic and Social Research Council (http://www.tlrp.org). His ongoing research interests include ontological questions regarding virtual reality and the nature of online teaching, Human memory, beliefs, attitudes and attitude change, and the potential of existential phenomenology as a framework for understanding human action and behaviour within both virtual and physical environments.

Dr. El Parker (el.parker@coventry.ac.uk) is a senior lecture in Engineering Geology and Programme Manager of undergraduate and research degrees in disaster management at Coventry University's School of Science and the Environment. She joined the University in 1999 after completing a PhD in Geology, which investigates patterns of climate change as evidenced in desert margin soils deposited since the last interglacial, at Liverpool University. Her other research interests include managing complex natural hazards in small Island developing States and the perception of risk and how this effects the development of response strategies. She has been involved in the development of online learning in Disaster Management since the institution-wide adoption of a VLE by Coventry University in 1999, and became particularly interested in developing online learning in Disaster Management while completing Coventry University's Postgraduate Certificate in Education.

Integrating Technology in Introductory Earth Science Classes

Neptune Srimal obtained his PhD in Geology at the University of Rochester. His research interests are mostly centered on the tectonic evolution of the Himalayas, where he spent considerable part of his life analyzing rocks and structures from very remote regions of this mountain belt. Srimal has taught undergraduate and graduate classes in SUNY, Buffalo and has been associated with FIU since 1992. He also was a Director of the Indian Geological Survey and thus has combined aspects of professional, academic and field life in his career which helps him in bringing real-world perspective in his class room lectures. He was involved with computers from the beginning of the PC era and has served as earth science database administrator and as webmaster. He teaches a number of traditional and online earth science courses at the Florida International University.

Integrating the Web into Environmental Decision Making Roleplay-Simulations

Professor Kirkpatrick, leads the strategic development of teaching and learning and the operational and strategic management of faculty development and flexible and distance learning at Monash University. She also provides strategic leadership for the Centre for Learning and Teaching Support. She has been Director of the Teaching and Learning Centre at the University of New England, and the Centre for Enhancing Learning and Teaching at Charles Sturt University. She has extensive experience in the field of faculty development and educational design and development. She has worked as a consultant for Australian universities, schools and government departments and won national and international awards for her teaching and educational design. Professor Kirkpatrick has a substantial record of publication and presentation in the field of e-learning and university teaching and is Chair of the Australasian Council on Open, Distance and E-learning.

Dr Robert McLaughlan is a Senior Lecturer in the Faculty of Engineering at the University of Technology, Sydney and Head of the Core Curricula Program. This program has a particular focus on professional and multidisciplinary studies within the undergraduate Engineering degree. He has a disciplinary background spanning geography, engineering and education. His educational research interests focus on frameworks for online role-based collaborative learning. These frameworks are structured around roleplay-simulation and debates. They have been adapted to various contexts involving engineering, science and social science students within the higher education sector. This work has won national and international teaching and learning awards and is used as an exemplar for a national project into Information and Communication Technology-Based Learning Designs.

Online Training for English as a Second/Foreign Language Teachers

Dr. Pamela L. Anderson-Mejías received her Ph.D. in Applied Linguistics/English Education from Indiana University in 1980. She has been working with teachers and teacher training for multicultural settings since 1977 at Indiana University, Southern Illinois University, the University of Hawaii, and the University of Texas—Pan American. She was a Fulbright lecturer in Tucumán, Argentina and Chile during 1989. Dr. Anderson-Mejías has published on issues in socio- and applied linguistics, as well as online instruction for the field, in journals such as the International Journal of the Sociology of Language, Hispania, Foreign Language Annals, English in Texas, SEDI Journal, the Internet TESL Journal, The Technology Source etc.

Using Geographic Information Systems (GIS) Software for Teaching and Learning

Gebelein received her B.A. in Biology from Syracuse University in 1994, her M.A. in Marine Affairs and Policy from the Rosenstiel School of Marine and Atmospheric Science (RSMAS)/University of Miami in 1996. She received her Ph.D. in Geography at the University of California, Santa Barbara, in 2001. She has been an Assistant professor of Geography at Florida International University since 2001. Professor Gebelein has been very active in proposal writing and has successfully received approximately $88,000 for a NASA Land Cover grant in 2001 to create land cover maps from satellite imagery of all the Greater and Lesser Antilles and $16,912.50 from the International Hurricane Center in 2003 to develop a new application for their Lidar instrument. She is also committed to excellence in teaching and regularly employs instructional technology in the classroom to facilitate learning. Dr. Gebelein is committed to seeking out new opportunities for students and has recently (May 2004) received $10,000 from the Kaufman foundation to support two students (1-undergraduate, 1-graduate) for several months to develop a business application using GIS and remote sensing technology.

Blended Learning and Geriatrics Education

Dr. Jorge G. Ruiz is assistant professor of medicine and director of the Geriatric Medicine Fellowship in the Division of Gerontology and Geriatric Medicine at the University of Miami School of Medicine. He formerly headed Geriatrics at Mount Sinai Medical Center of Greater Miami. Dr. Ruiz, a fellowship-trained geriatrician and a medical Informatics fellow of the National Library of Medicine, is board certified in internal medicine, geriatric medicine, and hospice and palliative care. An expert in e-learning in medical education, Dr. Ruiz is a founder and the director of GeriU (www.geriu.org), the Online Geriatrics University, a learning-management system for geriatrics education based at the Stein Gerontological Institute. He is the associate director for education/evaluation and an investigator at the Miami VA Geriatric Research, Education, and Clinical Center. His research interests include Web-based geriatrics education, development of reusable learning objects, and applications of blended learning in medical education.

Dr. Maria H. van Zuilen is assistant professor of medicine in the Division of Gerontology and Geriatric Medicine at the University of Miami School of Medicine. She received a diploma in geriatric nursing in the Netherlands and a PhD in clinical psychology from the University of Miami (UM). As co-director of the UM's 4-year longitudinal curriculum in geriatrics, palliative care, and pain management for undergraduate medical students and coordinator of the D. W. Reynolds educational grant for physician competency in geriatrics, Dr. van Zuilen has broad responsibility for medical school curriculum development, implementation, and evaluation. Together with the coauthors, she has developed a variety of preceptor-assisted and computer-assisted training materials in geriatrics for health-care professionals from different disciplines.

Tara Kai is an adjunct instructor of English at FIU and experienced with issues of content delivery via the Web and CD-ROM. She coordinates the D. W. Reynolds Project (Development of Physician Competency in Geriatric Syndromes) in the Division of Gerontology and Geriatric Medicine at the University of Miami School of Medicine. She is responsible for facilitating group discussion leading to consensus alteration of educational content and support and workshops on learning objects in distributed learning environments, needs assessment, and instructional design, production, and evaluation. She facilitates the precise steps of team members for adopting processes and methodologies in the development of exportable instructional materials for both preceptor-assisted and computer-assisted training and an exportable system for competency assessment. Ms. Kai also coaches division faculty in learning-content management system (LCMS) methods for the delivery of course content. Her first novel, Dar es Salaam (Bridge Works, NY), was published in October 2003. Her short stories have been published in U.S. and German anthologies.

Dr. Michael J. Mintzer, associate professor of medicine, has been with the University of Miami (UM) since 1985, holding a series of leadership positions in clinical medicine. He formerly served as director of the UM's General Medicine Private Practice and associate medical director and president of the Medical Staff for the UM Hospital and Clinics / Sylvester Comprehensive Cancer Center. In the Miami VAMC Dr. Mintzer first held the position of associate chief of staff for Geriatrics and Extended Care, followed by a tenure as medical center Chief of Staff, before assuming his present post as director of Community-Academic Partnerships in Geriatrics. As co-chair of the Statewide Steering Committee for Florida's Teaching Nursing Home Program based at the Stein Gerontological Institute, Dr. Mintzer's education and research focus encompasses geriatrics programs for the public, nurses, medical students, interns, residents, and faculty. He frequently lectures at national and international meetings and participates in the production and medical editing of a wide range of geriatrics educational materials.

Technology as Enhancement Tool for Professionalization and Cognition: The Effects of Ultra-Portability and Meta-Cognition Strategies

Dr. Mark D. Szuchman is Professor of Latin American History at Florida International University, where he also serves as Associate Dean for Faculty, Graduate Studies and Technology in the College of Arts and Sciences. Dr. Szuchman took his Bachelor of Arts degree in Latin American Studies at Brandeis University in 1969, and his doctorate in Latin American History in 1976 from The University of Texas at Austin. His research areas are urban and family history, and his publications have focused on nineteenth- and twentieth-century Argentina. He has received numerous grants in support of his research, which has resulted in the publication of several books and over 40 articles. His books include Mobility and Integration in Urban Argentina: Córdoba in the Liberal Era; Order, Family and Community in Buenos Aires, 1810-1860; The Middle Period in Latin America: Values, and Attitudes, 17th-19th Centuries; Revolution and Restoration: The Rearrangement of Power in Argentina, 1776-1860 (with J. Brown); and I Saw a City Invincible: Urban Portraits of Historical Latin America (with G. Joseph). His articles have appeared in major historical journals. He co-wrote, co-produced, and co-directed with Dr. Robert Levine Hotel Cuba: A Historical Diary of the Pre-Castro Jewish Experience. This documentary received the Hubert Herring Award for the best film of 1985 from the Pacific Coast Council on Latin American Studies. In his graduate seminars, Professor Szuchman has trained his students in the use of computers and computer applications, such as SPSS, for the purpose of demographic data analysis. More recently, he has incorporated technology and personal computing as part of the students' learning experiences.

Project Cine Math: An Adventure of Video Signals

Dr. George Kafkoulis is an associate professor of mathematics at Florida International University, in Miami, Florida.

Interactive Excel tutorials in a Quantitative Literacy Course for Humanities Students

Vera Frith is a lecturer and courseware developer in the Numeracy Centre of the Faculty of Higher Education Development at the University of Cape Town, South Africa. She came to this field via a Master's degree in Applied Science and a Higher Diploma in Education. Her work in the Numeracy Centre includes research into the assessment of students' quantitative literacy and curriculum and materials development for teaching students in a variety of different disciplines. She is particularly interested in the use of interactive Excel-based materials for mediating learning in Quantitative Literacy and Mathematics courses, particularly for disadvantaged students.

Jacob Jaftha is a lecturer in the Multimedia Education Group (MEG) at the University of Cape Town, South Africa. He is interested in the integration of computer technology in the teaching of quantitative sciences. His special interest is in the development of online tutorial learning environments that support interactive activities, shorten the feedback cycle and provide adaptive feedback. Before joining MEG, he taught mathematics at the University of the Western Cape for fourteen years where he first experimented with the introduction of computers in the mathematics classroom for teaching and research.

Robert Prince is a senior lecturer and coordinator of the Numeracy Centre in the Faculty of Higher Education Development at the University of Cape Town, South Africa. His work includes researching the assessment of appropriate quantitative literacy at the tertiary level, developing quantitative literacy interventions in various disciplines and at various levels across the tertiary curriculum, and the teaching and evaluation of these interventions. He is particularly interested in using writing and computer-mediated learning to improve students' mathematical literacy and mathematical concept acquisition.

Human Patient Simulators in Medical Students' Preclinical Curriculum

John L. Szarek is Professor and Chair of Pharmacology, and Director of the Simulator Program at Ross University School of Medicine in the Commonwealth of Dominica. He earned his Ph.D. in Pharmaceutical Sciences from the University of Kentucky College of Pharmacy, a B.S. in Pharmacy from the University of Illinois Chicago College of Pharmacy, and a B.S. in Biology from the University of Illinois Urbana. Prior to joining the Ross faculty, Dr. Szarek spent 16 years on the faculty at Marshall University School of Medicine. Much of his time at Marshall was spent in scientific research which resulted in numerous articles. Through his participation in workshops on teaching and learning and his own study, Dr. Szarek has developed and conducted workshops for medical school faculty on several topics. His interests focus on the use of technology in the classroom, classroom observations, microteaching, classroom assessment/research, and how students learn.

Ian Winston is an Assistant Professor in the Introduction to Clinical Medicine Department and Director of the Simulator Program at Ross University School of Medicine in the Commonwealth of Dominica. He earned his M.B.B.S. from the United Medical and Dental Schools of Guy's and St. Thomas' Hospitals in London. Prior to joining the Ross faculty, Dr. Winston was practicing medicine in the UK. He is responsible for the training of facilitators in the Simulator Program at Ross University School of Medicine. Dr. Winston's interests are focused upon family practice and psychiatry, small group learning in medical education, and the therapeutic effects of music.

The Virtual Student: Using Technology to Teach School Psychology

Ari S. Yares is currently a doctoral student in Temple University's School Psychology program. His research interests include the use of technology in psychology, training practices for school psychology, and the role of religion in mental health.

T. Chris Riley-Tillman is currently an Assistant Professor with the School Psychology program at Temple University. In his current position, Dr. Riley-Tillman provides training in intervention, consultation and assessment. Dr. Riley-Tillman's primary areas of research interest involve the study of the translation of research to practice, the analysis of behavioral consultation in reference to maximizing treatment integrity, and the creation and implementation of empirically based academic and behavioral interventions. Dr. Riley-Tillman currently serves as a reviewer for School Psychology Review, School Psychology Quarterly, and as guest co-editor of an upcoming special issue of Psychology in the Schools on transferring research to practice.

Catherine Fiorello received her Ph.D. in School Psychology from the University of Kentucky in December 1992. She also completed a clinical postdoc in ADHD at UK. She is currently an Associate Professor and program coordinator at the APA-accredited school psychology doctoral program and master's/certification program at Temple University. She specializes in assessment, especially developmental assessment in early childhood, assessment of children with disabilities, and cognitive and neuropsychological assessment. She has numerous publications and presentations in the areas of cognitive assessment, preschool assessment, LD, AD/HD, health-related disorders of children, inclusion of students with disabilities, and neuropsychology.

Marie C. McGrath is a doctoral student in Temple University's School Psychology Program. Her research interests include psychoeducational and neuropsychological assessment, school- and community-based intervention with at-risk youth, and training practices in school psychology.

From Blackboard to SMART Board: Training Techno-riffic Teachers

Dr. Jacqueline Hansen was born in the Rocky Mountains and raised in the Denver area. She earned a B.A. in Elementary Education from Doane College before becoming an elementary teacher in Grand Island, Nebraska. While she was teaching full-time, she earned an M.A. in Elementary Education, an M.A. in Educational Administration, and an Ed.D. in

Educational Administration, Curriculum and Instruction from the University of Nebraska. After 23 years of teaching children, Jacqueline left the elementary classroom to become an Assistant Professor of Education at Murray State University. Now she strives to teach pre-service and practicing teachers to love teaching as much as she does! Jacqueline creates education kits for the Special Stamps group in the US Postal Service. Her research interests include instructional technology, classroom management, and effective teaching strategies. Jacqueline and her husband, Allen, a computer specialist, enjoy singing, gardening, and riding their Harley Davidson motorcycle.

Utilizing WebQuests in a Preservice Teacher Education College Course

Robert Perkins is an associate professor and department chair for the Department of Foundations, Secondary and Special Education in the School of Education at the College of Charleston, SC. He teaches courses in instructional technology, special education and secondary education. He has a Bachelor of Arts in Political Science from the University of Massachusetts, a Master of Arts degree in Teaching in Special Education from West Virginia University and received his Doctor of Education in Curriculum and Instruction with a minor in Computer Education from West Virginia University.

Brenda Roughton has a Bachelor of Science in Early Childhood Education from Auburn University and a Master of Education in Early Childhood Education from the College of Charleston. She has taught first grade for two years, second grade for fourteen years, and third grade for six years. She is currently teaching kindergarten. In addition to teaching regular education classes, she has taught classes for the gifted and talented program. Ms. Roughton has presented at technology conferences on both the state and national level. She integrates technology into daily lessons and activities for her kindergarten class.

Integrating Technology into a Two Continent Consortium: Lessons Learned Teaching the TIRES Virtual Seminar

Sarah J. Mahler is an Associate Professor of Anthropology at Florida International University in Miami. She received her undergraduate degree from Amherst College and her graduate degrees from Columbia University. Her research and publications focus primarily on Latin American and Caribbean migration to the United States and the development of transnational ties between migrants and their home communities. Among her publications are American Dreaming: Immigrant Life on the Margins (Princeton 1995) and "Transnational Migration: Bringing Gender In" in International Migration Review (2003) co-authored with Patricia Pessar with whom she has also co-edited a special issue of the journal Identities: Global Studies in Culture and Power on gender relations in transnational spaces (2001). She has conducted research recently on religious praxis and its relationship to civic engagement among immigrants in Miami, including its transnational dimension, and on treatment for substance abuse provided by faith-based organizations serving Miami's immigrants.

About the Reviewers

John R. Cannon is an Associate Professor of Science Education at the University of Nevada, Reno, where he's taught undergraduate and graduate science education classes since 1992. His research interests include classroom learning environments, English-as-a-Second Language and science teaching, and teacher self-efficacy. Dr. Cannon is also the editor and publisher of the Electronic Journal of Science Education. In 2002, he authored a chapter on distance learning in science education in "Evaluation of Science and Technology Education at the Dawn of a New Millennium" (Kluwer).

Tony Erben is an Assistant Professor (FLE & ESOL) at the University of South Florida. He is past President of the Florida Foreign Language Association, Director of the Sunshine Coast Academic Alliance of Foreign Languages and Literatures as well as chair of the Florida Foreign Language Leadership Council. He received his Ph. D. in educational linguistics from Lancaster University in England, M.A. from the University of Sydney in Australia as well as his Graduate Diploma in Education and B.A. Before coming to USF, he was head of the School of Languages, Communication, and Art at the Central Queensland University as well as Director of the Centre for Language Immersion Teacher Education, where he established the world's first university degree program through Japanese immersion to graduate primary school teachers (LACITEP). His areas of professional specialisations lie in immersion education, sociocultural theory, critical literacy in FLE, as well as applications of technology in the FL classroom. Over this time, he has published numerous articles, chapters, and resources in the area of Foreign Language Education and ESOL.

Kinshuk is Associate Professor at the Massey University, New Zealand. He is also Director of Advanced Learning Technology Research Centre. He has been involved in large-scale research projects for adaptive educational environments and has published over 100 research papers in international refereed journals, conferences and book chapters. He is an active researcher in learning technologies and human computer interactions. He is currently chairing IEEE Technical Committee on Learning Technology, New Zealand chapter of ACM SIGCHI, and the International Forum of Educational Technology & Society. He is also editor of 'Educational Technology & Society' journal (ISSN 1436-4522) and 'Learning Technology Newsletter' (ISSN 1438-0625). He is editorial board member of more than a dozen journals and programme committee member of several conferences in educational technology area every year.

Delena Norris-Tull is an Assistant Professor in the Department of Education, University of Montana-Western. She specializes in science education and environmental education. She has published two books on field botany. Dr. Norris-Tull has taught courses on how to use computers and the Internet and self-published a booklet for teachers on how to use the Internet. She received a B.A. in Liberal Arts/English literature from the University of Texas, 1972, an M.S. in Fisheries and Wildlife/ Environmental Education, Michigan State University, 1975, and a Ph.D. in Science Education from the University of Texas, 1990.

About the Editors

M.O. Thirunarayanan is an associate professor of learning technologies in the College of Education at Florida International University (FIU), in Miami, Florida. He is currently also a Fellow of the Honors College at FIU. He earned his Ph.D. in 1990 from the College of Education at Arizona State University. Over the course of his career, he has taught science methods courses at the undergraduate and graduate levels, and learning technologies courses at the bachelor's, master's and doctoral levels. He was also instrumental in developing several courses and master's and doctoral level tracks in learning technologies.

Aixa Perez-Prado is an Assistant Professor in the College of Education at Florida International University (FIU), in the TESOL and Foreign Language Education programs. Her doctorate is from Florida State University in Tallahassee, Florida where she was a Bilingual/Bicultural Title VII Fellow in Social Science and Education. She also holds a Master of Arts degree in TESOL from Monterey Institute of International Studies in Monterey, California. Her research and writing has been primarily in the area of using online technologies in teacher education and with English language learners. She has taught English language learners and pre-service teachers both in the U.S. and abroad.

www.ingramcontent.com/pod-product-compliance
Lightning Source LLC
LaVergne TN
LVHW090306210926
848480LV00001B/151